ROUTLEDGE LIBRARY EDITIONS:
BRITISH IN INDIA

Volume 13

INDIA–CHINA
COMPARATIVE RESEARCH

INDIA–CHINA COMPARATIVE RESEARCH

Technology and Science for Development

Edited by
ERIK BAARK AND JON SIGURDSON

Routledge
Taylor & Francis Group
LONDON AND NEW YORK

First published in 1981 by Curzon Press Ltd

This edition first published in 2017
by Routledge
2 Park Square, Milton Park, Abingdon, Oxon OX14 4RN

and by Routledge
711 Third Avenue, New York, NY 10017

Routledge is an imprint of the Taylor & Francis Group, an informa business

British Library Cataloguing in Publication Data
A catalogue record for this book is available from the British Library

ISBN: 978-1-138-22929-7 (Set)
ISBN: 978-1-315-20179-5 (Set) (ebk)
ISBN: 978-1-138-73519-4 (Volume 13) (hbk)
ISBN: 978-1-315-23058-0 (Volume 13) (ebk)

Publisher's Note
The publisher has gone to great lengths to ensure the quality of this reprint but points out that some imperfections in the original copies may be apparent.

Disclaimer
The publisher has made every effort to trace copyright holders and would welcome correspondence from those they have been unable to trace.

STUDIES ON ASIAN TOPICS NO. 3

INDIA–CHINA COMPARATIVE RESEARCH

Technology and Science for Development

Edited by
ERIK BAARK and JON SIGURDSON

CURZON PRESS

SCANDINAVIAN INSTITUTE OF ASIAN STUDIES
Kejsergade 2, DK-1155 Copenhagen K

First published 1981

Curzon Press Ltd : London and Malmö

ISBN 0 7007 0138 9
ISSN 0142 6028

Printed in Great Britain by
Biddles Ltd, Guildford, Surrey

CONTENTS

CONTRIBUTORS

Aqueil Ahmad, Research Policy Institute, Lund, and Administrative Staff College of India, Hyderabad, India.

Erik Baark, Research Policy Institute, Lund, Sweden.

Richard Conroy, Science Policy Research Unit, University of Sussex, Brighton, England.

Björn Hettne, Department of Peace and Conflict Research, University of Gothenburg, Gothenburg, Sweden.

Ward Morehouse, Research Policy Institute, Lund, and Council of International and Public Affairs, New York, USA.

Jon Sigurdson, Research Policy Institute, Lund, Sweden.

PREFACE

China and India are continental countries with vast, still largely untapped, natural and human resources. Historically the cultures of these countries have played an important role in human development; there can be no doubt that, as their resources are fully mobilized for industrialization and modernization, the two countries will play an increasingly important role in world affairs.

So it is natural to find that comparisons of the two countries and their future roles in the global economic and political system fascinate people interested in international economics and the relative values of different political systems.

Comparisons can be carried out in many different ways: one may look at economic performance, development strategies, rural development, social structures, and many other aspects. One of these aspects, technology and science in development, has been especially studied at both the Scandinavian Institute of Asian Studies, in Copenhagen, and at the Research Policy Institute (previously Program) at the University of Lund. Against this background the two institutes decided in 1977 to convene a Scandinavian symposium on *Technology in Development: The Cases of India and China* in May 1978. The aim of the symposium was threefold:

1. To take stock of the present knowledge of mechanisms and processes of technology utilization in development in India and the People's Republic of China;

2. To shed light on the methodologies to be used for analysing technological developments in the two countries;

3. To help in the development of an analytical framework within which the experiences of India and China could become meaningful to each other as well as to other developing countries.

The focus of the symposium was further underlined by the fact that China's first National Science Conference was convened in March/April 1978, and that technology and science is receiving continued attention not only in India but in the world at large; for example at the United Nations Conference on Science and Technology for Development, convened in Vienna in August 1979.

Preface

The symposium brought together more than 30 people. Most were from Scandinavian countries, but there were also a number of specially invited foreign scholars and resident guest researchers. Papers of a comparative nature had been prepared by some of the participants, and these were read and discussed. The sessions became focused to a considerable degree on the availability of data, the comparability of data and questions of methodology.

In the sometimes heated debates suggestions were made that any analysis should focus on political conditions, social structure, centreperiphery dependency or empirical case studies. From this discussion one can discern a need to distinguish among three different modes of research: descriptive, analytical and normative approaches.

Descriptive studies are aimed at establishing typologies which classify issues and phenomena according to differences and similarities. Analytical studies are intended to provide a background for the creation of generalized theoretical models by successive testing of hypotheses through empirical research. In normative studies, the relative performance of various units - nations, regions, groups, etc. - are evaluated with reference to a normative standard. Each of these methods has particular fields of application and, if used in combination, they provide an efficient tool for comparative research.

In light of the very different views expressed at the symposium it was decided to summarize the proceedings in the somewhat unconventional manner which can be seen in this volume. Out of a total of ten papers presented at the symposium, only those of a comparative nature and outlook on methodological questions have been included here. In addition a major part of the volume consists of a methodological discussion of how to compare technology and science in India and China as well as an extensive survey of India-China comparative research. It is our hope that this will prove more valuable than a volume which would have included all the papers and a full verbatim report of the exchanges of ideas at the symposium.

As an outcome of the seminar a group of people had during 1979 formulated a research programme, *Technology and Culture,* for a comparative study of technological development in Asia and Western Europe. The programme is sponsored by the Committee for Future Oriented Research of the Swedish National Council for Research Coordination and Planning. It will focus on technical development and some of its conditioning factors in China, India, and Japan compared with the experience of such late industrializers in Europe as Sweden and Germany. It is our hope that the research programme planned for 1980-1983 will turn into an international collaborative effort with participating researchers from the countries mentioned.

I want to thank the Scandinavian Institute of Asian Studies, still my employer at the time of writing, and its director, Professor Søren Egerod, for the early decision to sponsor the symposium and to provide the necessary financial support for the publication of the present volume.

Donald B. Wagner, a research fellow at the Scandinavian Institute of Asian Studies, read large portions of the text and provided us with most valuable comments on the contents and the English.

Finally, I want to thank Mr. Erik Baark, research fellow at the Institute, who took on the burden of editing the volume, writing an extensive paper on methodological problems, and preparing the surveys of India-China comparisons and comparative studies in general. It is our hope that this volume will not only provide a stepping stone for researchers connected with our comparative work at the Institute, but also give others some valuable insights in the areas of comparative studies in general and China-India comparisons in particular.

Lund, December 1979.
Jon Sigurdson
Director, Research Policy Institute

STUDIES ON TECHNOLOGY AND SCIENCE IN INDIA AND CHINA: PROSPECTS AND METHODOLOGY

Erik Baark

This paper discusses the questions relating to the role of technology and science in development. As the symposium *Technology in Development: The Cases of India and China* addressed itself to these questions, it was revealed that comparative research on the experiences of India and China would present significant contributions on this issue. Therefore I will summarize here the proposals for future research areas presented at the symposium, and discuss theoretical and practical problems connected with comparative research on India and China.[1]

1. Technology and Science in Development

In the decades which have elapsed since the Second World War, researchers in social and political science have shown an increased interest in problems connected with development. Technology, in its conception as the systematic use of all technical knowledge, methods and operations in the control of nature, has been increasingly recognized as one of the most important elements in the development process. Technological modernization can justifiably be considered to be the most important element in the growth of developing economies, and the debate on development has thus increasingly centred on the role of technology in the modernization of developing economies.[2]

In addition, science has been recognized as a crucial factor in technological development. An advanced scientific establishment in developing countries appears to be one of the prerequisites for industrialization in the modern sense of the word. In most of the less developed countries, an independent scientific and technological base is lacking, and this problem can be seen as one of the major hindrances for modernization of the economies of these countries.

In the international debate on development, the role of technology and science in the development process has been the subject of a number of studies. These studies have approached the issue from a variety of conceptual frameworks and have suggested alternative solutions to the problems created by the introduction of modern technologies in developing countries. In this debate one can identify a number of central concepts which have been widely discussed during recent years.

The divergence between the socio-economic contexts in which modern technologies appear is the basis for one influential school of thought. By adopting the concept of *appropriate technologies*, the proponents of this school seek to call attention to the problems created by an uncritical transfer of technology from advanced countries to developing countries. Two problems are particularly noteworthy: first, capital-intensive, advanced technologies will often be inappropriate in the economies of developing countries, which lack the necessary capital; secondly, advanced technologies are usually not appropriate in the context of the labour-abundant developing economy. Labour-intensive technologies which do not demand large amounts of capital are therefore 'appropriate' in developing economies, and a more critical choice of technologies is proposed.

Another concept which has been widely discussed in recent times is *self-reliance*. When discussing this concept, certain development theorists have proposed that developing economies should go through a period of isolation from the international market in order to be able to establish a 'self-reliant' basis for technological development. The reason for proposing such a strategy can be found in the unequal economic conditions between advanced and less developed countries. This inequality has had the effect that developing countries often rely to a large extent on foreign enterprise in their own economies. In many cases this foreign influence, even if it may contribute to economic growth in the developing country as such, works in ways detrimental to the establishment of an indigenous base for technology and science. The self-reliant strategy of development is thus regarded as an effective means to counteract this foreign influence.

A third viewpoint which has become influential lately is that economic growth should be coupled with equity in order to be successful. In the concept of *technology for basic human needs*, the technology employed in developing countries should preferably be ordered to serve the needs of the population at large. The experience of many countries in the Third World indicates that roughly the opposite is the case: that economic and technological progress is used to benefit a limited, already privileged, stratum of society. Through the concept of technology for basic human needs, a strategy for technological and scientific development is proposed which could serve to alleviate the widespread poverty of the unprivileged strata of society.

Many of these new concepts have been applied in empirical research as well as in development programmes all over the world. Among the participants at the symposium, however, it was felt that comparative studies of India and China could provide a basis for further examination of the practical and theoretical implications of these concepts. One reason is that a number of the concepts mentioned above have been incorporated in the development strategies of both India and China. Another reason can be found in the similarities in macro-economic conditions of the two countries on the one hand, and the differences in approaches in development strategy on the other hand. An interesting aspect of the role of technology and science in development in India and China is the relationship between the particular social and political contexts of these countries and the development strategies chosen. The symposium led to the conclusion that India-China comparative research could be a valuable instrument for the attainment of a deeper understanding of the role of technology and science in development.

2. An Agenda for India-China Comparative Research

A selective bibliography was compiled in connection with the symposium where a list of India-China comparisons was included.[3] The compilers lamented the fact that 'literature on India-China comparative analysis is in short supply', and when a stock-taking survey of this literature was made for the present volume (see Chapter VI), it appeared that the field of India-China comparative research is biased in several respects.

The subjects most frequently treated in recent India-China comparative research appear to be comparative economic performance, studies in development strategies, and rural issues. When one notes that, in addition, most of this research has been carried out by Indian scholars, one is left with the impression that comparisons between India and China are made chiefly with reference to political issues in India. While this bias is obviously rooted in a genuine concern over the advantages of the Indian or the Chinese model of development, the purpose of the symposium was to call attention to a different aspect of India-China comparisons. This aspect is the promotion of a deeper understanding of the role of technology and science in development, not only in India and China but in all developing countries.

Therefore, a number of new research areas were proposed by the participants, and this section will attempt to delineate these research areas and their possible contributions to India-China comparative research.

a. *Political conditions*

Science and technology must be studied in the political contexts of both countries, as the fate of scientific and technological development is likely to depend on the policies adopted. Björn Hettne's paper on development strategies (included in this volume) provides an example of an analysis of political conditions. The paper compares the role of different political lines in the formulation of development strategies. Through his analysis of the dialectical evolution of policies on development issues in India and China, Hettne provides a detailed picture of the political conditions in which development has taken place.

While for instance Hettne's analysis gives an outline of the process of *policy formulation* in India and China, comparative studies of *policy implementation* appear to have been neglected so far. Particularly with reference to science policy, a discrepancy between the political goals formulated by governments and the actual implementation carried out by local authorities has presented serious problems in developing countries. Neither India nor China appear to be exceptions in this regard. Comparative research on the implementation of policies on technology and science in India and China is a vital element in the correct assessment of these policies.

b. *Social conditions*

The political contexts of science and technology strategies are embedded in the context of the social structures of both countries. In a comparative perspective, the social structures of a country may influence the technological and scientific developments. Comparative research on social structures in India and China can thus be considered essential for the specific studies of issues related to technology and science.

For example, a comparison of social structures within units of similar size in the two countries - e.g. villages - is a necessary prerequisite to studies on the effects of technology transfer. In the debate on the role of technology in development, the opinion has been voiced that the question of political and financial influence is most crucial for the successful introduction of modern technologies. One typical example is the differences which appear when modern agricultural technology is adopted by various rural social units. Some forms of agricultural machinery will be inaccessible to small farmers, but may be profitable to large landowners - and cooperatives organized by small farmers!

At the symposium it was suggested that the social context of technology and science in India and China could be analysed by means of comparative research on the *class structure* in these countries. This could incorporate a theory of the contradictions between social classes, a description of the ways in which social classes recognize and formulate their interests, and possibly a theory of class hegemony over state power. While the heterogenity of both societies must be recognized, comparative research on class structure and power relations in India and China is likely to facilitate an increased understanding of the social context for technological and scientific development.

c. *Economic conditions*

Social structures in India and China are relevant factors in the composition of a very important element in India-China comparisons: the economic conditions for development. In the experience of many developing countries, the introduction of advanced technology and science has placed severe strains on the national economy. India and China have the potential, as large nations, to devote considerable resources to technological and scientific development; but this potential may be exploited in different ways due to social and political factors.

One way to attack this complex problem is to make a comparative study of *capital accumulation* in India and China. The socio-economic structure of each country is likely to have a decisive influence on the pattern and effectiveness of capital accumulation in the economy. Thus, several programmes for social change in India and China have been motivated by economic efficiency. Units such as the People's Communes in China and Community

Development in India can be regarded as instruments to promote capital accumulation in the rural economy.

On the other hand, *allocation of resources* is an equally rewarding area for India-China comparative research. Policies on technological and scientific development are linked with the ability, within a developing economy, to allocate capital to this purpose. The establishment of scientific and educational institutions, for instance, depends to a considerable degree on available resources. A phenomenon like 'mass science' in China indicates that inexpensive, local scientific networks can be a solution to the problem of limited resources. The question of resource-allocation processes is also linked to the question of class interests mentioned above. In this area of research, a comparative study of the formulation and effects of economic planning is likely to present interesting results.

Finally, the questions of *employment* and *labour mobilization* are significant among the economic conditions for development. These questions seem to apply in particular to predominantly rural economies. Therefore, comparative research is essential on the Indian and Chinese experience with, for instance, labour-intensive technologies.

d. International conditions

The nature of world technological development is unfortunately such that the greater part of modern technology is situated in the industrialized countries. In addition, most of the prerequisites for further development of existing and new technologies are also found in these countries. Consequently a developing country is obliged to engage in a bargain on the world market for the technology it needs, and usually this bargain is conducted on the basis of conditions set by the industrialized countries. The effect is often that foreign influence on the developing economy becomes a major constraint to the independent formulation of a development policy.

India and China have approached this problem in entirely different ways during the last three decades. India has generally been more open to foreign investment, while China has insisted on varying degrees of isolation from the international market. In both countries, however, the concept of self-reliance has been frequently used in connection with development policies. In India, restrictions on foreign activities have been limited by an adherence to the principles of liberal capitalism and democracy. In China, on the contrary, self-reliance has at times implied virtual isolation from the world market. Although this subject has been treated in many comparisons between India and China, only a few of these have been devoted to the effects of these strategies on technology and science.

Finally, it was suggested at the symposium that comparative research on India and China as recipients or donors of technology would be valuable. The Soviet Union has, at different times, supplied technology to both China and India. If the effects of the introduction of Soviet technology are compared in India and China, this study could present an interesting case-study of India and China as recipients of technology. On the otherhand, both have supplied technology to Tanzania , and a comparative study of India and China as donors of technology might cast more light on an essential question: whether Indian or Chinese adaptations of Western technology might be more appropriate for developing economies generally.

e. Case studies

The preceding sections have outlined some of the important research areas in relation to conditioning factors. But suggestions were also made at the symposium regarding some case studies in technological and scientific areas where India-China comparisons might prove particularly valuable.

A paper by Richard Conroy included in this volume serves to illustrate the value of a case study in an aspect of technology in India and China. This paper analyses the development and performance of the industrial sector producing instruments for laboratories and factories in each country. It is concluded that the development of this science-based industrial sector has been significantly dependent on political and economical strategies in India and China. The study thus indicates the conditions for technological development, and the effects of these conditions, for a key industrial branch in the developing economy.

A number of research fields in technology and science will be outlined below; it should be noted, however, that these represent only a fraction of the potentially relevant study areas.

1. Education. The importance of education for research and development has often been recognized, and a few comparative studies of India and China have been devoted to this issue. However, most of this research has been concerned with higher education, training of scientists, and so on; some of the fundamental areas in education for widespread technical change have been largely neglected. This is the case, for instance, with vocational training for industrial workers, spare-time education through correspondence courses and television, and rural extension services. Such studies may reveal significant problems at the lower levels of the educational systems.

2. Energy. As a result of modernization in India and China, both countries have to face the problems of a smooth transformation from low levels of energy consumption to comparatively higher levels. In addition, a choice has to be made with regard to different technologies for production of energy, exploitation of natural resources, and reliance on domestic or foreign sources of energy. The experience of China and India in these areas may present alternative paths of development.

3. Military. Much of China's research and development activity has been applied in the defence industry as well as in the civil sectors. Indian defence has also rested on the level of technological and scientific research reached in the civil sectors of the economy. This relation between military and civil priorities can be observed in particular in the progress of high-energy physics. A case-study of this research field in India and China would probably bring forth interesting facts about the capabilities of developing economies in advanced science.

4. Medicine. The technologies used in the health sector - and in birth control - can be studied as a case of technology applied to social purposes. A comparison of the Indian and Chinese approaches to production of medical instruments, the pharmaceutical industry, and the application of traditional and modern medical techniques would be valuable in this connection. Furthermore, the contradictions between traditional and modern medical science in India and China, and their eventual integration to form a new medical tradition, will be worth studying as a case of science applied to social needs.

3. The Methodology of India-China Comparative Research

The cultural, political, and social differences between India and China imply that comparative research on issues related to technology and science will face a number of methodological problems. First of all, there is the problem of collecting material for the research. This problem includes questions such as the cultural and political bias of data, the comparability of two sets of data, and the scarcity or abundance of relevant data. The second problem is related to the analysis of the collected material. In this connection one must distinguish among a number of approaches by which the material can be arranged and employed in an analysis. Finally, the concept of methodology has come to include the viewpoints which the researcher may have on the purpose of the research, that is, the question of partisanship in research. Each of these aspects of methodology shall subsequently be discussed below.

The question of actual comparability of India and China was also discussed at the symposium. It was argued that differences in social and political structure in the two countries deprive comparative conclusions of their meaning. This argument, however, appears to be based on the biased approach of some forms of comparative studies; that is, the tendency to let India become the scapegoat for all mistakes in development policy. It should be pointed out that by explicitly choosing an approach among those described below, it should be possible to maintain a more neutral position.

a. The data

One matter that appeared to meet with universal agreement from the participants at the symposium concerned the reservations which must be made when data for comparative research on India and China is collected. The problems connected with data for India-China Comparisons have been adequately summarized by Ward Morehouse as 'jigsaw puzzle' scholarship in relation to Chinese sources, and 'needle in the haystack' scholarship in relation to sources from India. What these phrases describe is the nature of the process which the researcher must engage in when he is using Chinese and Indian sources. In the case of China, the information available is scarce and often constructed in such ways that they are not readily applied to the purposes of the research. The researcher is therefore obliged to collect separate pieces of data from a wide variety of sources and subsequently try to place these separate items in a pattern in order to construct a complete picture.[4] In the case of India, data is often available in staggering amounts, and the researcher will then have to sift out the relevant information from a large number of sources.

In the following short presentation of the major questions relating to data collection, however, I shall draw a distinction between qualitative and quantitative information, where qualitative information includes, e.g., political statements and descriptions of organization, while quantitative information refers to, e.g., statistics on volume of production, prices, etc.

One of the major problems is the *availability* of data. As pointed out in the preceding discussion, this problem is most acute with information from China. In general, however, the availability of quantitative data is more restricted than that of qualitative data. The Chinese in fact release many statements of a political nature, and when for instance the general political aims of the Chinese leadership in relation to economic planning are analysed, there are a number of sources which can provide a substantial amount of information on the organizations involved, the measures taken to implement the planning objectives, and so on. The collection of quantitative data, such as statistics on growth of production, is unfortunately often impeded by the Chinese habit of releasing such information sporadically. In addition, such data is frequently formulated in percentages of increase and decrease in relation to years for which actual figures are also missing. The Indian statistical service, by contrast, is organized according to the scope and concepts of Western counterparts. A wealth of statistical information is thus available on the national and regional level of the Indian economy and society.

With reference to the internal and external circulation of information, China can consequently be characterized as a 'closed system', while India is an 'open system'. The result of China's closed system for information availability has been that estimates, in part based on data from the 1950s, have often been the only way to describe China's performance in recent times. A change has been initiated in the last few years, but this change has not yet resulted in an amount of quantitative information comparable to that of India. The open system of India has enabled researchers to have access to much larger amounts of data, both quantitative and qualitative, but several authorities have indicated that this fact may conceal some biases which are nevertheless inherent in the information.

Consequently, the *reliability* of the available data is an additional question which the researcher must consider. It has been quite common among researchers to question the reliability of Chinese data, but the absence of alternative sources in relation to which the data can be checked will often have the effect that the information must be used as it stands. But the reliability of data can depend on several factors; some forms of information, notably qualitative data on political issues, may be unreliable on account of editing, omissions, changes in scope of concepts, etc.; other forms of information, e.g., quantitative data on production may not be reliable simply because the source is itself unaware of the actual figures.

Another problem, which appears to apply equally to the Chinese and the Indian contexts, is the lack of a continuous relationship between macro-level information and micro-level information. A typical example is that figures for agricultural mechanization may be available for villages or communes, and perhaps also for the national level, but absent for an intermediate, regional level. A comprehensive framework for the interpretation of the micro-level figures may thus be lacking. The problem is, however, also part of general methodological problems concerning the focus of comparative research, and it will be taken up again in the following section.

Once the researcher has found ways of dealing with problems of availability and reliability for the data from India and China, the difficult question of assessing the *comparability* of data must be considered. This question will often have to be dealt with through specific research into the nature of the data used. An illustrative example of differences in the nature of quantitative data is the fact that figures for production of food-grains in Chinese sources include potatoes and unhusked grain, while Indian statistics exclude potatoes and are based on the weight of grains after husking. Obviously such figures must be corrected before a comparison can be made. But problems of comparability are not restricted to quantitative data. Differences in concepts used in qualitative data may be even more annoying in comparative research. The term scientist, for instance, carries different implications in Indian and Chinese sources. These differences refer not only to the educational qualifications of Indian and Chinese scientists, but they apply equally to divergent conceptions of the role of scientists in society, their status in relation to other groups, the purpose of research generally, and so on. Consequently, the information released by Chinese and Indian sources regarding, e.g., the number of scientists graduated per year must be carefully studied before a comparison is made.

In conclusion it must be emphasized that it is an essential task for anyone engaged in comparative research to make sure that a critical evaluation of the reliability and comparability of data has been conducted throughout the process of investigation. Furthermore, the results of such a critical evaluation should be made explicit when the research is presented in published form.

b. Focus of research

It was stressed several times during the symposium that comparative studies of India and China would be possible on several levels and with various foci. The comparisons could be made on both macro- and micro-levels and an intermediate level was also suggested. In addition, the comparisons could focus on synchronic as well as diachronic comparisons.

Comparisons on the macro-level are exemplified by such sweeping studies as the development strategies or science policies in India and China. Research on the intermediate level could be studies of, for example, industrial sectors or provincial regions in both nations. At the micro-level, one could imagine research on the patterns of production in factories or villages. There is no doubt a great deal of heterogenity in both nations, and regional, cultural and political differences in various sections of both nations in the vertical as well as the horizontal perspective, and this heterogeneity must be taken into account when the subjects for India-China comparisons are chosen. The choice of level of research has also important implications for both the availability and the comparability of data, as well as for the best approach in research methods. There is a constant interaction between the availability of data, the comparability of the data obtained, and the focus of the study.

In an intermediate-level study, for instance, the information obtained will often have to be illuminated in macro-level perspective in order to determine the comparability of the two sets of data; by implication, a small set of data may be sufficient for a comparison to be made if information relating to other levels are used to qualify the data with respect to comparability.

The choice of focus will also influence the choice of methodological approach. In the following section I shall present an outline of three possible types of approach; each of these approaches may have strong points of its own when the level of research has already been chosen with the subject. The analytic approach, for instance, involves the formulation of a hypothesis which is to be examined with different sets of data. This approach is less likely to be valuable at the macro-level - where a descriptive approach may yield the best results - but would probably be an adequate tool at the micro-level.

With reference to the question of synchronic versus diachronic studies, it was suggested at the symposium that the synchronic analysis which is usually applied in India-China comparisons may be too limited in many subjects. An example may be found in the Chinese and Indian efforts to transform the mode of production in agriculture, which have taken place in different historical periods; in this case, a diachronic focus of research may be more valuable.

c. Approaches

Comparative research can be approached through different research methods. In the following exposition, it has been considered useful to distinguish between three forms of approach: the descriptive, the normative, and the analytic research method. The establishment of these concepts has had a dual purpose; first, the distinction between the three different approaches have been employed in the classification of research presented in Chapter VI; secondly, each of the three approaches has particular applications according to the subjects studied, the focus of research, and so on.

The methodology of comparative research is, however, also the subject of several theoretical works which have been published in recent years. Much of this theoretical research has been used as a basis for the conceptual framework for research methods outlined in this paragraph, but in order to introduce researchers to this interesting subject, an annotated bibliography has been compiled for this volume.[5] It should be emphatically stressed that a more conscious approach to these theoretical questions would benefit the field of India-China comparative research.

Descriptive research methods. The concept of descriptive research methods is here used to designate research where the main objective is a description of similarities or differences in performance, approaches, and so on. Such a description of similarities can eventually provide a background for the inclusion of various phenomena in existing cross-national typologies. The establishment of typologies is recognized as one of the basic elements in comparative research; when only two countries are compared, however, the establishment of typologies assume less importance. The description of similarities and differences then becomes valuable from another point of view. The value of descriptive studies lies in their contribution to other forms of comparative research, notably analytic research methods. Thus, the formulation of hypotheses, which is the initial phase of analytic research, is facilitated by descriptive studies. It has been suggested, for instance, in a preceding paragraph that descriptive comparative studies of, for example, class structure and power relations should be undertaken in order to provide a background for other comparative studies.

One must be aware, however, that descriptive comparisons do not lead to explanations in the scientific sense of the word. Scientific explanation is usually associated with generalized theoretical statements on the influence of one or several factors on the outcome of a process. Descriptive research is not able to present such theoretical statements. For example, the purest form of descriptive comparisons is presented in anthologies of papers on specific subjects in India and China.[6] Such anthologies illustrate how a descriptive comparison often leaves it to the reader to make conclusions on the influence of, for example, politics or society on the course of development.

Normative research methods. The concept of normative research comprise studies where the performance of one or both countries is compared with a normative standard. The normative research conducted with reference to India and China can be roughly divided into two categories. One category has been the comparison of the economic performance of India and China with respect to an explicit or implicit set of norms, generally derived from studies of advanced economies.

These normative studies in economic performance appear to be undertaken primarily to indicate the success of a political ideology, e.g., socialism or liberalism, in attaining Western standards of development.[7] The application of such norms to developing economies may be questioned, and the value of these normative studies in economic performance has been further diminished by the uncertainty of the data used.

Another form of normative research is offered by studies in which the performances of India and China are considered with respect to a proposed strategy for development. In this case, the researcher defines an ideal way to handle a specific problem, e.g., self-reliance in industrial production, and then proceeds to investigate to what degree the performances of India and China approximates this ideal strategy. The paper by Ward Morehouse included in this volume, for instance, evaluates the Chinese and Indian experience in creating a local technology system. This study may indicate, on the one hand, deficiencies in the theoretical model for a local technology system, while, on the other hand, improvements in either the Chinese or the Indian strategy may be proposed.

Finally, normative research can be conducted as a comparison of the performance of each country *vis-à-vis* its own goals, for instance as formulated in five-year plans.

Analytic research methods. The concept of analytic research is based on the objective of such research to reach generalized theoretical statements through testing of hypotheses in empirical research. It is in analytic research that the social scientist finds something comparable to the controlled laboratory experiment of the natural scientist. In the chemical laboratory, it is possible to add one substance to another in a test tube, to heat to a specific temperature, and to observe and measure the reaction. But it is impossible for the social scientist to add or to remove a social class in a society in order to examine the effect of this class upon political change. In other words, the social scientist is rarely able to manipulate the variables directly. However, by testing a hypothesis in comparative, cross-national research, he can manipulate the experimental variables indirectly.[8]

The basic methodological procedure of analytic research thus starts with the identification of a process which may be influenced by a number of variables. Obviously, the process must have taken place in both countries. The researcher can then proceed to formulate a hypothesis indicating that one or several factors have had a significant influence on the process in question. Finally, this hypothesis should then be tested through observation of the effects of various factors on the process, based on the empirical data collected.

There are several difficult problems connected with this approach. One is the intricate relationship between the data available and the possibility of formulating hypotheses which are likely to be confirmed or rejected through the empirical research. This problem can best be solved through a critical evaluation of the data with respect to availability, reliability, and comparability. Thus, there is a mutual influence between the collection and evaluation of data and the formulation of hypotheses for analytic research. Another problem is the influence of the choice of focus of research. For example, research on micro-level phenomena requiring personal access to information on the lowest levels of society may be impossible in China, and hypotheses which must be tested in this area are likely to prove futile.

It must nevertheless be emphasized that however difficult the analytic research method may appear, it is probably the only way to improve general theoretical knowledge about development processes. The reason for this is that the testing of hypotheses envisaged in analytic research is a fundamental element in the formation of scientific paradigms. Therefore, if a general theoretical understanding of the role of technology and science in development should emerge in the future, this theory would have to rest on empirically examined paradigms; these empirical investigations are probably best conducted through analytical comparative research, for instance into the Indian and Chinese experiences.

d. Partisanship

In the preceding paragraphs, problems of methodology facing anybody engaged in comparative research on India and China has been discussed. These problems must be dealt with by the researcher irrespective of his partisan viewpoints on the purpose of the research conducted. But partisanship in research will always be explicitly or implicitly present in the researcher's selection of topic, data and research method.

In research on the role of technology and science in development, and in India-China comparative research in particular, the question of partisanship is unavoidable. The reasons for this are manifold. First, the problems of development are intimately connected to political issues, and research on development issues are, at the same time, highly relevant for political decisions. Therefore, the conclusions reached in research on development issues will probably, consciously or otherwise, serve the political purposes of a class in society, a nationalist cause, or another political entity. Secondly, the cultural and ideological contexts of the Indian and Chinese society are opposed to each other in a way which suggests a partisan choice on behalf of the researcher. Thirdly, the roles of India and China in international politics are dependent on their approach to development, and the researcher may find partisanship unavoidable in studies of these approaches.

Having so far argued that partisanship is unavoidable in the research field explored in this volume, it still remains to propose the limits and nature of such partisanship. The British historian E.J. Hobsbawm has proposed that

> partisanship in science (using the word in the general sense of the German *Wissenschaft*) does not rest on disagreement about verified facts, but about their selection and combination, and about what may be inferred from them. It takes for granted non-controversial procedures for verifying or falsifying evidence, and non-controversial procedures of argument about it.[9]

Thus, partisanship does not imply a methodological choice with respect to the procedures for scientific research, but rather implies a conscious choice of topic and working hypotheses. This point is important because it underlines two essential conditions which should be respected regardless of partisan positions. First, that it would be beneficial for a limited research field, like the India-China comparative research treated in this volume, if the results of research undertaken from different partisan points of view would be mutually relevant. This effect, it has been argued above, can be gained by following research methods for comparative research outlined in this section. Secondly, that the methodological approach and the partisan position of the researcher should be explicitly stated in published research.

While these limitations in the scope of partisanship are designed to reduce negative effects of such partisanship, the advantages of partisanship must also be emphasized. Partisanship properly applied advances science because it provides an incentive to change the terms of scientific debate, a mechanism for injecting new topics, new questions, etc., from outside. For example, an analysis of the influence of class structure on technological developments in India and China, undertaken from a Marxist point of view, may promote a critical evaluation of this subject. Likewise, studies based on a partisan approach rooted in an Asian cultural outlook would certainly challange the euro-centric conception of technological development.

NOTES

1. Suggestions and proposals enter this paper at various points, but the form and contents of it remain my own responsibility. I wish, however, to express here my gratitude to all participants for their inspiration.
2. Notably in connection with the 1979 United Nations Conference on Science and Technology for Development.
3. Aqueil Ahmad and Robert C. Hsu, *India and China: Technology & Science in Development. A selective bibliography*, Lund 1978, pp. 43-46.
4. Michel Oksenberg, 'Sources and Methodological Problems in the Study of Contemporary China' in A. Doak Barnett, ed., *Chinese Communist Politics in Action*, Seattle & London 1969, pp. 577-606.
5. See Chapter VII.
6. As in Kuan-I Chen and J.S. Uppal, (eds.) *Comparative Development of India and China*, New York 1971.
7. An illustrative case is Subramanian Swamy, *Economic Growth in China and India: A Comparative Appraisal*, Chicago 1973.
8. For a further elucidation of this argument, and a longer treatment of analytical research methods (*the* comparative method, according to the authors), see Robert T. Holt and John E. Turner, 'The Methodology of Comparative Research', in Robert T. Holt and John E. Turner, ed., *The Methodology of Comparative Research*, New York 1970, pp. 1-20.
9. E.J. Hobsbawm, '"Partisanship" and the Sciences' in *Culture, Science et Développement. Contribution à une Histoire de l'Homme*, Toulouse 1979, p. 268. Hobsbawm's paper contains a brilliant discussion of the subject, which I have relied heavily on in my own exposition.

SELF-RELIANCE VERSUS MODERNIZATION : THE DIALECTICS OF INDIAN AND CHINESE DEVELOPMENT STRATEGIES

Björn Hettne

Introduction

Why comparative studies of India and China? Let us first consider the changing concerns of this early established field. In the fifties and early sixties India and China were often regarded as two great Asian experiments, one democratic, the other totalitarian, which could serve as models for other developing countries. Their growth record was the undisputed yardstick showing the viability and effectiveness of the two systems.

From the mid-sixties and until the death of Mao Zedong in 1976 for different reasons the two experiments became increasingly incomparable. One reason was that the Chinese government had stopped releasing statistical figures in 1959. Another was that China, due to a combination of lack of information and an over-enthusiastic response from the 'new left' in the West, was thought to be moving toward some kind of utopia. At the same time India entered a period of economic stagnation and recurrent political crises ending up with the Emergency of 1975-6, which demolished the last argument in India's favour: the democratic facade.

We have now, as it seems, reached a third phase of comparative studies of India and China. After the 1977 election friends of India again raise their voices stating that, after all, the strength of the democratic system has been amply demonstrated. Simultaneously the present rulers of China are denouncing the Maoist experiment as a 'lost decade', while facing a spontaneous movement calling for a fifth modernization: democracy. This new situation will most probably modify the assumptions involved in comparative studies of the two countries but it is too early to say exactly in what way. Optimistically, the 'competitive', approach will never be revived; it is also doubtful if an overall comparison of economic performance is of much value. The differences in historical background, in resource endowment, in climate and ecology are so great that such an exercise would hardly be very rewarding. More fruitful comparisons could for example deal with specific economic sectors or organisational structures, where the major biases can be eliminated. We would also suggest that general thematic comparisons, where historical and other differences enter the analysis as explanatory factors, should be made to a larger extent. This has been attempted in the present essay, where we are exploring the ideas of 'self-reliance' versus 'modernization' in the context of Indian and Chinese history.

In the contemporary discussion on development strategies, particularly in the context of the New International Economic Order, one often comes across these two fundamentally different competing perspectives. We do not see them as incompatible, although those who are idealogically committed to one or the other may feel so. Rather they are conceived as dialectically interrelated, which implies that they have a corrective effect on each other.

According to the *modernization*-perspective, development in the Third World is necessarily a repetition of the historical experience of contemporary industrialized countries. This Western world view is shared by Liberalism and Marxism alike.[1] The most naive expressions of this development paradigm have, it is true, been demolished; but the basic notions, particularly as regards technology in relation to modern industry, remain, for obvious reasons, very relevant. It may be argued that technology is somehow more 'universal' than for example a political institution, but we feel nevertheless that the form and content of technology is so deeply influenced by a specific culture that a certain technology also implies a certain social structure. Thus, 'modernization' will here be used more or less as a synonym of 'Westernization', and it is important that this specific meaning of the term is kept in mind.

The other perspective is less imitative and more indigenous; it conceives the real meaning of development as a progressive change and improvement of given structures rather than development through 'creative destruction'. *Self-reliance*, as opposed to both dependence and imitation, is a crucial concept within this tradition. The ideology of *local* self-reliance in particular, emphasizing rural development, is commonly but not universally referred to as 'populism'. One reason to be careful with this word is that it has different connotations in Asia and Africa on the one hand and Latin America[2] on the other. However, the concept is hard to avoid, as it in our view constitutes a 'third dimension' in development thinking; it questions the primacy of technology and industrialism, so characteristic of both Marxist and Liberal development thinking, and stresses instead the human potential. In stressing the need for a maximum of local self-reliance and in attempts to solve the rural-urban contradiction, populism, furthermore, rejects the notion of economics of scale and indeed *all* large-scale and centralized solutions. It is our contention that 'Maoism' and 'Gandhism' to a large extent embody these populist ideas in China and India, and that they therefore constitute two similar albeit far from identical, models of development. In both countries they serve as alternatives to the more conventional modernization paradigm.

In both India and China the self-reliance strategy and the modernization strategy have been dialectically related to each other largely in way in which they affect the social power structure. This relationship is shown in Figure 1.

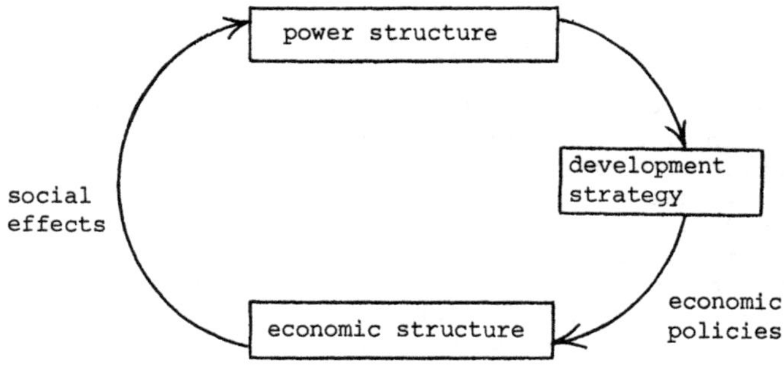

FIGURE 1

Thus, in order to explain why a particular development strategy is adopted, one must account for the power structure and the way various established institutions and groups attempt to promote their own interests. Certain economic policies will, on the other hand, affect the power structure in such a way that the economic structure is changed as a result. This also implies a change in the relationship between social groups. Thus there will always be a continuous process of adaptation between the development strategy and the power structure, resulting in more or less dramatic 'swings' in development policies. This further implies that there may be a difference between the formulation of a strategy and its actual implementation. The more heterogeneous the power structure, the greater this difference will be, as is amply illustrated by the cases of India and China.

1. Historical Foundations of Development Thinking

1.1 The Quest for Modernization

As emphasized above, 'modernization' implies a comparison between two units; and the development perspective associated with this concept is that the unit which is found to be 'old-fashioned' should try to catch up with the unit which is more 'advanced'. Because of the international power-structure created by imperialism, this comparison, along with the obvious conclusion to 'modernize', was more or less forced upon the non-European world.[3] The choice was between modernization and destruction by imperialist powers. The latter was the inescapable fate for small-scale, isolated and technologically 'primitive' societies. On the other hand, the great agrarian civilizations, with more developed political superstructures, where the challenge could be assessed and some remedy be thought of, had a possibility of escaping destruction. Japan is of course the great and so far only example of a society which accepted the challenge, modernized, and was able to beat the imperialist countries on their own terms.

Both India and China were deeply affected by Western imperialism and the response to the Western challenge has a long and complex history. Since India was colonized at a very early stage (before industrialism and modern imperialism), she never had the opportunity to follow the Japanese example.

The case of China lies somewhere between those of India and Japan, as far as the imperialist challenge is concerned. Unlike India, China was never colonized in the strict sense of the word, although a number of imperialist powers carried out their policies with very little regard to the Chinese government. Unlike Japan, China was therefore unable to avoid imperialist control. The Self-strengthening Movement in the nineteenth century was too weak to follow the Japanese example. One of the reasons for its weakness is that it was opposed by the Imperial Court; another is that the early modernizers were themselves very ambivalent in their approach to modernization. In the introduction we mentioned that technology was inseparable from culture and society. This fact was not quite realized by the early 'self-strengtheners' of China. They were traditional Confucian scholars advocating 'Chinese learning for the substance, Western learning for practical application.'[4] They entertained the hope that Western military technology might be mastered by skilled artisans and craftsmen, so that the Chinese *literati* could devote themselves to more important things. It is significant that the emergence of 'Western studies' was conceived mainly as an alternative for those whose path to an official career had been blocked.[5]

A less ambivalent attitude toward modernization and Westernization was held by the leaders of the May Fourth Movement (1917-21) the purpose of which was to maintain the existence and independence of China as a nation. To the young intellectuals who were in the lead it had become clear that conservatism and traditionalism (Confucianism) now had to be scrapped altogether. The worship of the old was replaced by enthusiasm of the new, i.e. Western knowledge and ideas.[6] As we shall see, this instrumental approach to Westernization and modernization dies hard in China.

In India the cultural conflict between Western-type modernization and various forms of neo-traditionalism has perhaps a longer and more violent history than in China. As a fully colonized country India has felt the British cultural impact since the middle of the nineteenth century, when a radical modernization policy was started (particularly under Dalhousie 1848-56). The Mutiny of 1857 can be seen as a traditionalist and revivalist response to that policy. This uprising was subdued, however, and the later struggle against colonialism operated from a very different social base and with a very different intellectual content. As an instrument of this struggle the Indian National Congress was formed in 1885. Most leaders of this organisation were influenced by Western constitutional ideas, and as the emphasis of the agitation gradually turned from political to economic issues, the British were held responsible for the poverty of India. This was the thrust of Dadabhai Naoroji's celebrated 'drain theory'.[7] Significantly it was not, however, British rule but the 'un-British' rule that had created the poverty in India. There was nothing wrong with the liberal principles of Britain. The problem was rather that the British in India did not adhere to them.

The ideals of modernization thus became part of the freedom struggle where they competed with Gandhi's ideals of self-reliance and Hindu revival. The former conquered the Indian modern elite, while the latter were more instrumental in mobilizing the masses. This dualism has characterized the Congress party ever since. In the thirties there was furthermore, a split among the modernizers between those who preferred capitalist methods and those who believed in socialism and, more specifically, in the Soviet model of development. Much of the political controversy in India has occurred between these two ways of modernization, whereas the 'Gandhists' have been rather passive since the death of Gandhi in 1948 and until recently. We shall return to this matter below.

1.2 The Roots of Self-reliance Thinking

Whereas the quest for modernization by definition implies an effort at catching up with 'advanced' nations, self-reliance is, although normally inspired by an external challenge, a more indigenous kind of development thinking, and therefore more difficult to account for. The most simple approach is, of course, to look at the two great personalities which personify self-reliance thinking in India and China, Mahatma Gandhi and Mao Zedong, but it should then be remembered that each belongs to a specific ideological tradition.

India and China are old, proud, and self-conscious civilizations, and it is only to be expected that self-reliance thinking should be particularly strong and elaborated here. Possibly the strongest tradition in this respect is that of China, which regarded the rest of the world as 'barbarian' and only through force could be made to trade with the West. The traditional Chinese attitude to foreigners is clearly expressed in the famous edict of Qian Long to the King of England in 1793, from which the following passage is quoted.

> As to the request made in your memorial, O King, to send one of your nationals to stay at the Celestial Court to take care of your country's trade with China, this is not in harmony with the state system of our dynasty and will definitely not be permitted. Traditionally people of the European nations who wished to render some service under the Celestial Court have been permitted to come to the capital. But after their arrival they are obliged to wear Chinese court costumes, are placed in a certain residence, and are never allowed to return to their own countries. This is the established rule of the Celestial Dynasty with which presumably you, O King, are familiar. Now you, O King, wish to send one of your nationals to live

> in the capital, but he is not like the Europeans, who come to Peking as Chinese employees, live there and never return home again, nor can he be allowed to go and come and maintain any correspondence. This is indeed a useless undertaking . . . As a matter of fact, the virtue and prestige of the Celestial Dynasty having spread far and wide, the kings of the myriad nations come by land and sea with all sorts of precious things. Consequently there is nothing we lack, as your principal envoy and others have themselves observed. We have never set much store on strange or ingenious objects, nor do we need any more of your country's manufactures . . .[8]

This attitude goes a long way to explain why the Chinese 'modernizers', as discussed above, had a hard time. It is obvious that Western ideas became weapons in the struggle for power within China, and that there were some good political reasons for suppressing those who held it necessary for China to Westernize in order to survive. It is significant that the 'modernizers' who took power after the death of Mao used similar arguments, the only difference being that it is now the Soviet Union that figures as the main threat against Chinese survival. It has been pointed out that Mao comes from that Chinese tradition which is continental and agricultural - a tradition which sees China as a series of self-sufficient rural units, united by a central bureaucracy. This tradition is essentially anti-urban, anti-foreign, and anti-commercial. The Yanan-syndrome was basically a reemergence of this tradition, in a revolutionary mould. But new elements were added.

Gandhi's ideas on economic development were not derived exclusively (or perhaps in this case not even primarily) from the Hindu tradition; he was in fact heavily influenced by the Western, especially Russian populist tradition. Like the economics of populism, Gandhian economic thought can only be understood as a *counter-theory* to Western economics. From Tolstoy, Gandhi borrowed the concepts of egalitarianism, simplicity, and asceticism (the last two of course having their Hindu parallels), from Ruskin the emphasis on ethics in economic development, and from Kropotkin a hatred of economic centralization and bureaucracy.[9] Thus, the ideal for Gandhi was the rural economy. He visualized the society as being built up of circles of self-supporting and self-contained units, exchanging only necessary commodities with other villages, where they were not locally producible. Agriculture and handicraft based strictly on subsistence, in place of an economy of surplus, would be the foundations. Land should be communal property, and the implements and tools of a cottage industry should belong to the family traditionally engaged in it. Since, in this society, owners of the means of production would also be labourers, exploitation based on individual property would be impossible.[10]

Mao, like Gandhi, was deeply rooted in the soil of his country, but in his case too the Western impact seems to have had a catalyzing effect on his ideas. Mao is often pictured as a Marxist-Leninist, but it is of importance to point out that the influence of Marxism came rather late in his life. Like Gandhi he was early influenced by the anarchist and populist thinking of Kropotkin, Bakunin and Tolstoy.[11] In fact the first great Chinese Marxist, Li Dazhao, who had a great impact on young Mao, seems to have been equally influenced by Russian Narodism and Marxism.[12] Even if he never directly studied the Narodnik writings he took a strikingly similar stand in emphasizing the village community and the need for the intellectuals to 'go to the people'. His Marxism, like later that of Mao, was therefore of a very unorthodox kind.

There is no denying that Mao was a Marxist. He himself said he was, so this is not the issue. However, as has been convincingly argued by Maurice Meisner, the relationship between Marxism-Leninism and Maoism is a rather ambiguous one.[13] According to Meisner the influence of Marxism-Leninism never completely overwhelmed the populist strain in China and, furthermore, a powerful populist impulse was to become an integral component of the Maoist version of Marxism. One of the fundamental characteristics of Maoism is the still unresolved tension between Leninist-type elitism and the populist belief that the peasantry possesses an innate socialist consciousness.[14] In the case of Gandhi we find the populist ideology in its more pure form.[15] It could be argued that what differentiates Maoism from Gandhism is Marxism and that what Mao and Gandhi have in common is populism.[16] The common elements in Gandhism and Maoism have lately become more widely acknowledged and comparisons between them are no longer rare.[17] At least the Soviet Marxist characterizations of Gandhi and Mao are more or less the same, both being described as protagonists of petty bourgeois nationalism and voluntarism.

Comparisons are always highly suspect, since two phenomena may seem different or similar depending upon what aspects are considered. Here we are more concerned with similarities than differences, and it is particularly the problem of self-reliance which is in focus.

Galtung has observed that self-reliance primarily belongs to the realm of psycho-politics.[18] Gandhi and Mao gave the ordinary Indian and Chinese self-respect, dignity, and the capacity to defend themselves.[19] One of the first concerns of both men was with the image of weakness projected by imperialism on India and China.[20] Both Gandhi and Mao realized the 'modernity of tradition', i.e. they were innovative traditionalists. Gandhi often referred to *Ram Raj*, and Mao spoke of *Da tong;* these were the two utopias in the classical Indian and Chinese traditions. Neither Mao nor Gandhi believed that material incentive could be a prime mover for individual effort and efficiency. Both were anti-elitist in the classical populist sense and both wanted to narrow the distinctions between physical and manual labour, between cities and countryside, and between workers and peasants.[21] Gandhism and Maoism are also alike in the emphasis on simple living and identity with the poorest. The loincloth and the 'Yanan way of life' are symbols that have different historical settings but are similar in substance.[22] Finally, the communes of Mao and the village communities of Gandhi (stripped of the differences in ideological approaches) practically amount to the same thing. According to Sethi, this is one of the most remarkable points of similarity between the two.[23]

2. *The Power Structures of India and China*

The main reasons for the different emphases in the Indian and Chinese development strategies should of course be looked for in the different *power structures* of the two countries. By this term we refer to the relative importance of various classes and institutions in the decision-making process (and their changes over time), rather than the more formal distribution of power as expressed in the constitutions. We shall confine ourselves to those components of the power structures which are identified in the simplified model[24] in Figure 2.

This model (Figure 2) is no substitute for class analysis but it identifies more or less institutionalized spheres of interests which acquire meaning only insofar as they are related to the class structure of society. This can, however, be only crudely attempted here.

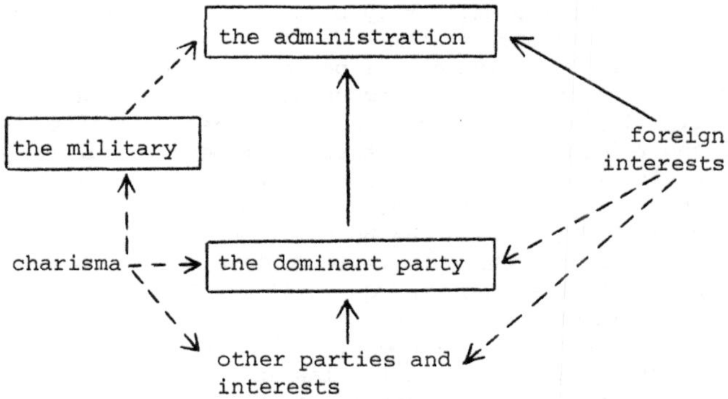

FIGURE 2

The *dominant parties* in India and China have been the Congress Party and the Communist Party. The Congress Party had suffered a great setback in 1967, split in 1969 and was almost demolished in 1977. It has, in terms of voters, a middle-class, worker, and peasant base, but its leadership has all along been dominated by a Westernized elite, of which one segment has had capitalist and another, much smaller one, socialist leanings. This heterogeneous and somewhat paradoxical structure should be seen against the background of the party's long history as a liberation movement. The mass mobilization against imperialist penetration never developed toward social revolution as it happened in China, but relied instead on Gandhi's charisma as well as local patronage structures, which of course were not very revolutionary. The orthodox character of the Indian Communist movement further facilitated this reformist strategy.

The recent breakdown of the Congress Party should not be overdramatized. It is very likely that the party may be restored. The Janata party, which took over as the 'dominant party', has functioned much in the same way, represented essentially the same interests, and was in fact partially based on segments of the disintegrating Congress Party.[25] The difference between Janata and Congress is thus one of emphasis within a ruling coalition of classes and elite groups. During the Emergency (1975-6) segments of the industrial bourgeoisie, had the upper hand, whereas the rural rich, particularly the 'kulak', elements, occupied a corresponding position in the Janata government. However, Janata was never a homogeneous political force. It was rather a delicately balanced and continuously changing alliance system.

After a less successful phase of organizational work in the urban areas the Communist Party of China became primarily a peasant movement, mobilized by the issue of land reform and by the rather unorthodox (populist-inspired) policy of the party under Mao's leadership. Immediately after the revolution the recruitment policy again shifted toward the bourgeoisie and the intellectuals, and the characteristic dualism (red versus expert) took shape. The Cultural Revolution (1966-9), which almost destroyed the party, was intended to reduce the 'expert' element. Since the army was instrumental in maintaining law and order, a conflict between the army (Lin Biao) and the party (Zhou Enlai) became unavoidable. The party was obviously the more successful actor and has now been re-established as the dominant political force in China - together with the bureaucracy.

The *administration* is of course formally in charge of executing political decisions but it also has a certain autonomy. Both India and China have long and complex bureaucratic traditions. In India the colonial administration was taken over almost intact, and only the local administration (Panchayati Raj) is a later innovation. There has always been a tension between politicians and administrators, though this does not exclude a certain symbiotic relationship. During the Emergency (1975-6) an 'administrative society', similar to the political system of the colonial era, was re-established, but this was not of long duration. The administrative structure in India is and has always been extremely centralized and the bureaucrats of course have a stake in this system. The same is the case in China, where a clear tension exists between the mandarin tradition on the one hand and the anti-elitism manifested in Maoism on the other. After Mao there has been a most forceful revival of bureaucratic power.

Charisma is perhaps a controversial concept, but somehow one has to account for the remarkable power to influence the flow of events that has been exercised by exceptional leaders like Mao and Gandhi. Of course their 'charismatic power' was not equally influential on *all* groups and categories but should be explained by reference to cultural traditions and class interests.[26] Charisma reflects a *social relationship* rather than a psychological quality, and the strength of charisma is therefore explained by the expectations raised by a certain leader in his 'followers'. The character of charisma is determined by cultural factors, such as the stress upon moral purity and self-denial in Hindu tradition or heroism and struggle in Chinese tradition. It is difficult to see how the cultural Revolution would have been possible without Mao, or the demolishing of the Congress Party without Jayaprakash Narayan, applying a basically Gandhian technique of political change.

The *military* has so far been of direct political importance only in China, but it has been intimated that Indira Gandhi after the catastrophic 1977 election actually asked the military to take over.[27] The Indian military is extremely professionalized, however, and does not seem to harbour political ambitions. One student of the Indian army notes that 'possibly no other group in south Asian society is so critical of politicians on particular issues, and yet is so strong in its support of the political system.'[28] The contrast with the Chinese army is striking. The People's Liberation Army (PLA) has been both a political and military organization, besides being a production force, all in line with the Yanan way, and enforced by Mao against the 'professionals'. The Korean War boosted the latter view, as did Soviet aid to the specialized branches, dependent as they are on modern technology. The Great Leap and Cultural Revolution, on the other hand, reinforced 'guerillaism'. This, however, has again changed after the demise of Lin Biao, and in the future the army will also be 'modernized'.[29]

Foreign interests, finally, is admittedly a somewhat cryptic term but here we refer primarily to other nations which by diplomatic and other means attempt to influence the internal policies of a certain country. In the case of India these interests have, as we shall see, been quite formidable in spite of India's rather independent foreign policy. Typically they have operated through the mechanisms of financial dependence in periods of foreign exchange constraints. In the case of China, the policy of self-reliance implied a break with the Soviet Union, and it is an open question whether the modernization strategy recently adopted will lead to a new dependence. If so, the dependence will not be upon the Soviet Union.

Now, to summarize the implications for 'self-reliance' versus 'modernization' of the Indian and Chinese power structures there is, particularly in the

Indian case, a heavy bias in favour of modernization. Initially there was a cleavage between the Congress Party and the bureaucracy, and between the bureaucracy and the business class (the former mainly Brahmins, the latter merchant castes and minorities); but gradually there have emerged close ties between party and business, while the cleavages have been bridged, thus creating a solid establishment bent on industrialization and modernization. Already in 1956 (at the height of 'socialism') it was clear that the public sector versus the private sector was a minor issue, and that most important business concerns willingly accepted government participation in management without any fear of vigorous extension of governmental activity for example in the field of heavy industry. The struggle between 'capitalism' and 'socialism' has therefore been a matter of emphasis rather than principle and both ideologies share the same modernization paradigm. To the 'modernization forces' must be added the military, whose modernization was emphatically stressed after the 1962 debacle against the Chinese, and whose increased strength became clearly visible in the war against Pakistan in 1971. The Gandhian elements, on the other hand, rapidly declined in political importance after the death of Gandhi. His selected heir, Vinoba Bhave, emphasized the metaphysical aspects of Gandhism rather than the practical-political aspects, and in doing so he led the Gandhian movement into the desert.

Jayaprakash Narayan (in India known as J P) much later rediscovered the political struggle component and, now also supported by the bulk of the Sarvodayaists, achieved a remarkable revitalization on Gandhism which put an end to the rule of the Congress.

In China too the strength of self-reliance has been primarily connected with the charismatic factor; but in contrast with Gandhi, Mao lived long enough to create a very distinctive political tradition. In this he evidently had the support of the old core of the army, who continued to believe in the Yanan way, as well as those many students who came from non-academic backgrounds and therefore rejected the elitism of the universities. In the peasants, Mao of course also had a rather secure base. On the other hand he was often opposed by the urban interests, proletarian and non-proletarian, as well as the more orthodox communists within the party, the bureaucracy at large, and parts of the military, eagerly waiting for modernization. This makes the balance between modernization and self-reliance a little more even in China than in India.

However, in both cases the odds against a radical policy of indigenous development and self-reliance must ultimately be considered as rather heavy.

3. *The Dialectics of Development*

After the above presentation of the two contending views on development, and having discussed, in a sketchy way, the power-structures in India and China, it should now be possible to put things together and see how the opposites interact. This will give concrete form to the development process in the two countries. It has been fairly common to give a dialectically inspired exposition of the changes in development strategy in China, whereas this approach is less common in the case of India.[30] The reason for this may be that those who analyse the development in China may be influenced by Mao's own dialectic view of things, but it is also very clear that the oscillations are more pronounced in the case of China, where the alternative lines have been more clearcut. In India the politically relevant alternatives have been 'capitalist' versus 'socialist' modernization, whereas the emergence of the JP-movement in the early seventies was never fully articulated as an alternative development path. Gandhian policies, as for example the Community Development programme, have been emphasized now and then, but never as part of a consistent development strategy. Thus the pendulum movement in the Indian case is somewhat distorted by the fact that there is a triangle of policies, 'capitalism', 'socialism', 'Gandhism', which operates within the framework of a 'reconciliation system'. On the other hand in China we have the more dualistic contradiction between 'Maoism' and 'revisionism', which operates within the framework of a 'mobilization system'.[31] These terms are here used in a contextual way, which means that they have limited meaning outside a concrete Indian or Chinese situation. 'Modernization' and 'self-reliance', as defined above, are on the other hand here used as universal concepts which may serve our comparative purposes. Thus in India 'capitalism' and 'socialism' represent two varieties of modernization, while 'Gandhism' is the Indian form of self-reliance thinking. In China 'revisionism' represents modernisation and 'Maoism' self-reliance.

Let us now define the dialectics of development, first for India and then for China.[32]

3.1 India: Capitalism, Socialism and Gandhism

Even before independence the three development paths of Capitalism, Socialism and Gandhism were widely discussed in India. Three plans containing these diverse approaches were presented in the forties: the Bombay Plan, expressing the visions of the industrialists, the People's Plan, emanating from the socialist camp, and the Gandhi Plan, written by disciples of Gandhi.

In this way the heterogeneity of the Congress was clearly reflected in the development debate. When the First Plan was announced it became clear that the capitalist forces led by Sardar Patel had been the strongest. This plan was extremely pragmatic and may be described as a combination of Capitalist and Gandhian elements, the most conspicuous example of the latter being the Community Development Programme. However, there was very little Gandhian about the *implementation* of this programme, which was a completely bureaucratic exercise.

Only during the Second Five Year Plan (1956-61) was it possible to speak of a consistent development strategy. The plan had been formulated by P.C. Mahalanobis, inspired by the Soviet Model. This implied emphasis on large industries and public ownership, i.e. modernization in the socialist way. Even in agriculture a 'socialist pattern' of economic development was stated as a goal in the Nagpur resolution passed by the Congress in 1959.

This shift to the left had been facilitated by the death of Sardar Patel, by the successful elections in 1951-2, and by the encouraging economic development during the First Plan. It is also very likely that Nehru's visit to China in 1954 had played its part. All this contributed to a strengthening of the hands of Nehru, who in turn left much of the responsibility of the planning process with Mahalanobis. The approach in the plan was, however, extremely theoretical, and reflected current ideas on the basic importance of capital accumulation as the *sine qua non* of development. In international finance circles the radicalism of the Second Plan was never much appreciated, and when India in the late fifties lived through an acute payment crisis, the World Bank and other agencies of international capitalism were able to exercise a certain influence over the economic policies of the Indian government.

The Second Plan failed miserably in mobilizing resources for the ambitious investment and growth targets. On the whole the planning exercise in India has been quite unrealistic, and after the Second Plan, which may be considered the only 'real' plan, the tendency has been a ritualization of the planning exercise.

The Third Plan (1961-6) was basically a repetition of the general objectives of the Second Plan, but turned out to be even less successful. In the mid-sixties there was a decisive shift to the right in economic policy; this can be seen most clearly in the sphere of agriculture. The Green Revolution meant that Gandhian agrarian policies were given up altogether. The emphasis was now on productivity rather than the creation of self-reliant village communities, and the organizational model of this modernization policy was clearly the capitalist firm. The Nagpur resolution was quickly forgotten. The reasons for this shift to the right were: the death of Nehru and the rise of the Syndicate; the political organization of Indian conservatism in the Swantra party; the reduction of federal power; monopolism in the private sector; and, again, an increasing influence of foreign (US) interests, as shown for example in the devaluation of the rupee in 1966.

The shift in development policies had important social and political consequences. In the late sixties the social tensions both in the countryside and in the cities increased dramatically (Naxalism, strikes, riots etc.). A crisis within the Congress led to a split (Congress-O and Congress-R) and the rise of Indira Gandhi in the elections of 1971 and 1972 to a position of power comparable to that of her father. At that time many expected a straightening-up of the political system under the banner of 'socialism'. India approached the Soviet Union, and Congress (R) received full support from the Communist Party of India (CPI), but in practice very little changed. Indira Gandhi's great alliance was based on political 'outsiders' who were given an opportunity to replace the old State bosses, who formed the Syndicate, with the result that corruption reached new heights. The Fifth Plan was launched under great controversy regarding its degree of realism. Whatever realism there was, it disappeared during the oil-crisis of 1973-4. After 1971 the Green Revolution also lost momentum and agricultural production began to fall; this contributed to the price rise. All this led to a political crisis in 1973-5, which saw the growth of the JP-movement (a Gandhian movement led by Jayaprakash Narayan) and ended with the Emergency of 1975 and 1976.

The JP-movement, which started in early 1974 in Bihar, was the first serious challenge that the Congress regime had faced since independence. It represented a resurgence of Gandhian political activism after more than 25 years of passivity. As mentioned above this passivity was caused by the particular interpretation of Gandhism made by Vinoba Bhave, but in the early 1970s many Sarvodayites, led by JP, challenged Bhave's views. The 'constructive approach' was, they felt, not sufficient. *Political struggle* was also necessary.[33]

The Sarvodaya organization was, however, too weak (except perhaps in Bihar) to carry out JP's 'total revolution', and therefore JP made use of the organized political opposition to attain the necessary infrastructure. The Gandhian element in the JP-movement became more and more diluted as the movement was elevated from its original Bihar context and transformed into an all-India movement. In spite of this the political establishment was alarmed, and on 26 June 1975 the President of India declared a state of national emergency due to 'internal disturbances'. Though the Emergency was the result of an acute political crisis, there were of course underlying economic causes, which in a summary way may be described as a contradiction between the transactional form of political system ('politics of distribution') and the stagnating economy. A more authoritarian approach was therefore felt to be necessary, since there was now less to distribute. The Emergency was not a very sudden change however. It had been preceded by nearly a decade of increasing authoritarianism and police violence.

During the Emergency Indira Gandhi made desperate efforts to get the general political and economic crisis under control. Thousands of political opponents were imprisoned and a 20-point programme was launched to solve the economic problems. This was adhocism (which may be said to have started with Indira Gandhi's 'Stray Notes on the Economy' at the 1969 Bangalore meeting of the Congress Party) in its most pronounced form. There was no consistency whatsoever in the programme, but it implied a straightening-up in the administration, which led to a tension between certain officials and the rural elite.[34] On balance the Emergency was more beneficial to the industrial than to agricultural interests. The enforcement of the family planning programme (not included in the 20 points), and the rough way in which it was implemented, backfired on the Government, which was already in a process of disintegration caused by the activities of the Indian 'Gang of Four' (Sanjay Gandhi and his associates).

The reasons for Indira Gandhi's decision to hold elections in 1977 are still rather obscure. She may have been unaware of the real state of public opinion, or she may have found it healthy to dissociate herself from the 'Gang of Four'. In any case the election destroyed the Indian dominant party, which in itself was a most remarkable outcome. A 'new' party emerged *after* the election to take over government responsibility. This organization, the Janata Party, was a most unusual and peculiar political configuration, and the most unbelievable thing about it was perhaps that a *Gandhian* economic programme was announced by the new government. This needs some clarification.

As discussed above, the Indian political scene in 1974 and 1975 was dominated by the JP-movement. A distinction should be made between, on the one hand, the movement as such, which was more or less Gandhian, was mainly student-based, and relied on the Sarvodaya organization for its core leadership and, on the other hand, the greater political alliance surrounding it. This alliance was composed of most of the opposition parties, from the Communist Party of India (Marxist), whose attitude was rather ambivalent, to Congress (O), with the Socialists and the Jan Sangh constituting the more enthusiastic core. (The Indian Socialists maintain certain Gandhian traditions, and the Jan Sangh had recently gone through a face-lift which made its outlook rather Gandhian too). Another party, Bharatiya Lok Dal, was mainly a North Indian 'kulak' party. Most of Janata's political base was, incidentally, North Indian, and in the South, even in the great debacle of 1977, Congress fared rather well.

What kind of economic programme was the new government pushing? Janata's election manifesto had spoken out against 'elitism, consumerism, and urbanism'. Light industry was preferred to heavy industry and rural to urban. It was, furthermore, necessary to steer away from both capitalism and socialism. A typical passage read:

> Cities have their place, but, if rural India cannot provide economic opportunity and creative outlets for the growing masses in the countryside, we will be forced along the capital-intensive, urban-oriented, and centralized path of development followed in the West. The experience of the last

> three decades has only underlined the relevance and validity of the values that Gandhiji placed before us.[35]

After the election the Janata government's economic declaration turned out to be much more vague:

> In the economic sphere the government is pledged to the removal of destitution within a definite time-frame of 10 years. Relative neglect of the rural sector has created a dangerous imbalance in the economy leading to migration of people from rural areas to urban centres. The farmer has been denied reasonable and fair prices for his products. Allocations for agriculture and related developments have been grossly inadequate and the need to improve conditions in the villages has received scarce attention. More than a lakh of villages do not even have the most elementary facilities for drinking water. My government will follow an employment-oriented strategy in which primacy will be given to the development of agriculture, agro-industries, small and cottage industries especially in rural areas. High priority will also be given to the provisions of minimum needs in rural areas and to integrated rural development. To the extent possible at this point of time, the fifth five year plan will be reviewed. The planning process will be revitalized and work on the sixth five year plan will be taken up without delay. My government will announce at the time of the presentation of the final budget later this year the details of the economic programme that is proposed to be followed.[36]

In early 1978 the Government released the draft 'sixth plan' (1978-83), a new kind of *rolling plan*, which differed from previous plans in that it should be continuously revised. In presenting the plan in Lok Sabha the Prime Minister said that it was 'necessary to go on assessing the implementation of the plan from year to year in order that shortfalls in implementation can be corrected and effective attention given to priorities'.[37] The emphasis on agriculture (43 per cent of total outlay) and rural industries was maintained in this document; it was described by the opposition as 'most retrograde and reactionary', as it 'threatened to reverse the process of industrialization and take the nation back to the colonial era'. The Prime Minister explained that the government's aim was to reduce dependence on outside help and achieve self-sufficiency. He also stressed that in the past the plans remained unimplemented to a large extent because of lack of involvement of the people. Now the government would organize panchayats and municipal ties on a 'partyless basis' to ensure involvement of 'all people' in the implementation of schemes for agriculture, cottage industries, and other rural developments.

The bold programme of self-reliance incontrovertibly *implies* a fundamental change in the development strategy. The heavy odds against its implementation stem not only from the lack of unity within the government but also from various *structural* constraints. The most obvious constraint is that the existing income distribution is not compatible with a re-orientation towards a decentralized domestic production. Second, it is not possible to allocate production quotas to large-scale, small-scale, and cottage industries without substantial state interference. Third, the discussion of 'appropriate technology' has overlooked the time-lags involved in infusing adequate organizational, managerial, and entrepreneurial inputs into the labour-intensive sectors.[38] Fourth, the bureaucratic structure of Indian society will possibly make the fate of the new panchayat policy similar to that of the Community Development in the 1950s.

Let us now turn to the *political* constraints. In order to assess the realism of that kind of economic programme it is necessary to consider the political power structure at the time. This is summarized in Figure 3.

It should be emphasized that the figure only summarizes the institutional conflict pattern at the national level. Below the elusive surface of formal organizations there is an immensely complex social infrastructure of caste conflicts, class-struggle, religious tensions, and clashing economic interests, which differs in its concrete manifestations from one State to another. The power game that goes on in the superstructure is ultimately determined by this changing infrastructure, but the scope of the present essay does not allow an elaboration of these connections.

Even with regard to superstructures the model is much too simplified but a few comments will add the necessary nuances. To start with the Janata corner of the conflict triangle, the lifetime of the Janata government was a little more than two years. During these two years its energy was spent on a power struggle in which the leading parts were played by the prime minister

Moraji Desai (former leader of Congress-O) and Charan Singh (leader of the Jat-Kulak party BLD).

The Political Cleavage Pattern in India 1978-1979

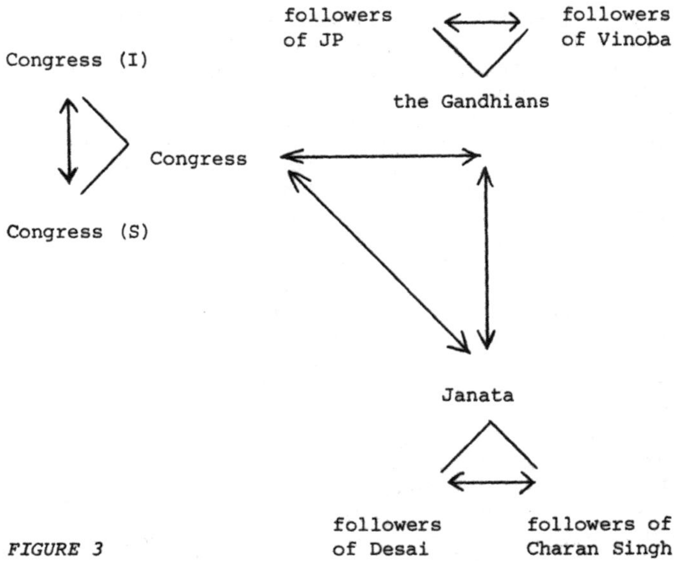

FIGURE 3

The major factor keeping Janata together was the threat coming from Indira Gandhi. The Congress party had been split into Congress (I) and Congress (S), led by Swaran Singh, on the issue of her leadership. Indira Gandhi had used the comfortable position of Congress in the southern State of Karnataka as a base, but in July 1979 the Karnataka prime minister Devaraj Urs challenged her leadership, which resulted in still another 'congress party', Congress (U).[39]

As the third corner of the triangle we see the 'Gandhian' establishment dominated by Jayaprakash Narayan. The Gandhian activists in India are a nebulous category which includes Sarvodaya workers engaged in the various activities which have their origin in Gandhi's Constructive Programme. They number about 50,000 persons, attached to over 8,000 Gandhian institutions.[40] We have earlier referred to the split in the Sarvodaya movement between followers of JP and followers of Vinoba Bhave. It is necessary to make a distinction between the JP-movement and the Janata party in spite of the charismatic role of JP in forging the alliance that crushed the dominant party of India. JP and his followers became more and more disillusioned with the Janata government and from time to time JP spoke of starting a new movement, but his state of health did not permit this. In an interview with Geoffrey Ostergaard in January 1979 JP said:[41]

> It was another historical opportunity lost . . . I don't know what new initiatives are called for or are necessary. But some new initiatives have to be taken by somebody . . . I'm out of the picture now because of my illness.

It is obvious that the Gandhian image of the Janata government to a very large extent was a concession to the role played by JP in the struggle against Indira Gandhi.

Even if the government had wanted to implement a Gandhian economic programme it was too paralysed by internal conflicts to accomplish much. Already in the summer of 1978 it was very close to breaking down, as Charan Singh was forced by Desai to leave the cabinet, only to return as Deputy Prime Minister in early 1979. By summer the tensions burst out again, and this time it was Desai who had to resign after a wave of defections from the Janata Party. Charan Singh then took over the prime minister post but had to resign in August after Indira Gandhi had withdrawn her support.[42] Once again the 'old lady' turned out to be 'the crucial element around whom the whole political process is revolving'.[43]

Indira Gandhi got what she wanted. A mid-term poll, expected to take place sometime in early 1979, was announced by the President. A desperate power game in order to reorganise the large number of factions into viable political alternatives started. In the midst of this unprecedented political disintegration JP died, physically broken by his term in prison during the Emergency and psychologically broken by disillusionment from watching the performance of what mainly had been his creation, the Janata party. Was that the end of a self-reliant and indigenous approach to India's economic and social ills?

3.2 China: Maoism and Revisionism

Initially the Chinese development strategy was little more than a copy of the Soviet model of development. It did not work in China, however. The reason was that the process of socialist accumulation had much earlier reached the limit of mass starvation in China than in the Soviet Union, and that many Chinese Communists, notably Mao, were not prepared to pay the political price that Stalin had paid. Thus emerged the uniquely Chinese (and Maoist) Great Leap Forward in 1958. This development strategy, which included the organization of People's Communes, clearly expressed Mao's visions about the future Chinese society and the means to achieve it. There is a close correspondence between this strategy and the 'Yanan way of life', the society created by the Communists in the Liberated Areas during the forties. It immediately shook the relations with the Russians, who in 1960 discontinued their aid, thereby contributing to a great leap - backward.

From 1959 when Mao found it necessary to leave the presidentship to Liu Shaoqi, his position was weakened, and in the early 1960s a completely new development strategy emerged. It was not a return to the Soviet model since, for example, agriculture rather than heavy industry was emphasized. However, Mao's critics were anxious to minimize the conflict with the Soviet Union. The populist emphasis on the supreme knowledge of the people was rejected, and the Party began to stress the crucial role of the intellectuals and the professionals in the development process. It was no longer considered essential for scholars to work in fields and factories. Rather they were encouraged to concentrate on studies. The hierarchical system of education was reintroduced at Peking University. Encouraged by the President, Liu Shaoqi, and the Party Secretary, Deng Xiaoping, intellectuals severely criticized Mao's economic policies. These critics were mainly authors and artists, and the critique against Mao was consequently given an allegorical form. This is no doubt one reason why Mao's counter attack was a Cultural Revolution. Later this concept was incorporated into the Marxist doctrine as one of the more significant theoretical innovations of modern times.

Since Liu Shaoqi was Mao's foremost opponent during the Cultural Revolution, a comparison between the two is called for.[44] In spite of the outright repudiation of everything that Liu allegedly stood for, which accompanied his purge in the Cultural Revolution, there are reasons to believe that their differences were perhaps more in political style and method than in revolutionary goals. This is at least suggested by the fact that there is nothing which indicates that Liu did not believe in the Great Leap, though he in retrospect admitted that it had been a failure, which Mao never did. However, the methods used by the radicals in the Cultural Revolution were more than he could accept, since they implied the destruction of the party and the administrative machinery. In short *Liu could never accept, nor could he understand the populism of the Cultural Revolution.* He had not experienced the mass campaigns in the 'red areas' and having been responsible for the political work in the 'white areas' during the war, he had had to rely on very different methods. He concluded that 'empty "revolutionary" shouts, which have no practical meaning, must never be made'. The two varieties of communist experience in China - in the White and the Red areas - explain much of the two subsequent 'fronts' of the party: the bureaucratized, urban-oriented, technocratic, and Western-influenced group versus the more egalitarian, populist and more local-oriented group. Liu decidedly belonged to the former. Liu, in accordance with established Marxist tradition, concentrated on mobilizing workers. When he spoke of 'the masses' he meant *workers*, while Mao meant *peasants*. In a 1941 essay (*The Class Character of Man*) he spoke of the peasants' 'lax ways, conservatism, narrowmindedness, backwardness' in contrast with the 'solidarity, mutual co-operation, sense of organization, and discipline' of the urban proletariat.[45]

The personality types of Mao and Liu could not have been more different, but of course this in itself is no reason why they should have become political enemies. However, in an era of 'destructive creation', these differences were bound to develop into contradictory positions, a struggle between 'two lines'.

The Cultural Revolution was a revitalization of Maoism. Assisted by the army, Mao cut the bureaucracy to pieces and all but crushed the Communist Party. In terms of the Chinese power structure it was a strengthening of the military element. In terms of regional balance of power it was a strengthening of the Shanghai group. In terms of Marxist philosophy it pointed to the need for a continuous revolution in the superstructure of society, where there was a permanent danger of capitalist revivalism which could affect the socialist infrastructure as well. In terms of development strategy it was a re-emergence of populism and the indigenous Maoist model of development.

This model contains six important components. As its core we see the elimination of 'three unequal relationships': between *city and countryside*, between *industry and agriculture*, and between *intellectuals and manual labourers*.

Mao conceived these relationships as contradictions to be resolved in a dialectic process, resulting in completely new economic, social and political structures.

The model is furthermore based on the principles of *self-reliance*, *participation* (*mobilization*), and *decentralization*. Self-reliance, the antithesis of 'dependence', is one of the key concepts in the recent discussion on development and underdevelopment. It is not wrong to say that Mao was one of the great popularizers of the concept. It should not be interpreted as 'autarchy', but implies a high degree of reliance on one's own resources and, very important, autonomous goal-setting and decision-making. As we discussed above, self-reliance thinking has a long tradition in China.

Participation and mobilization have been characteristic features of Chinese economic policies under Maoist inspiration. It is not correct to evaluate these features primarily as *means* to development. They should also be regarded as *ends* in themselves, reflecting the Maoist emphasis on equality and popular involvement. The principle of decentralization is an application of the same values to regions and levels of administration.

This model was thus basically similar to that of the Great Leap Forward. How did it work out this time? In retrospect it looks as if the Cultural Revolution, like its predecessor the Great Leap Forward, had adverse effects on production. This was recognized already in 1968 in a frank speech by Zhou Enlai. Whereas the Great Leap Forward suffered its worst setbacks in the agricultural sector, it was now industrial production which was most severely affected. The Shanghai workers in particular had been extremely active during the Cultural Revolution, though their activism was more due to their own grievances than to any commitment to the revolutionary goals. The disruptive effects on industrial production reached a peak in 1967, and the level of production both in 1967 and 1968 was lower than in 1966. Only in 1970 was the situation normalized.

In contrast to the case of the Great Leap Forward the economic policies of the Cultural Revolution did not imply any great structural changes. The People's Communes were largely intact, and in spite of recurrent campaigns against material incentives the private plot was left undisturbed. The most radical changes seem to have taken place in the field of education, the stronghold of Chinese elitist tradition. Theoretical studies were now downgraded in favour of practical training. Each university was required to establish workshops and agricultural plots. Whereas the failure of the Great Leap Forward brought the expansion of small-scale industry to a halt, the Cultural Revolution was characterized by an increased emphasis on the development of a local self-reliant industry. The functions of this industry included the utilization of agricultural raw materials, the training of a large rural industrial work force and the production of iron, cement, coal, and machinery required by the agricultural sector.[46] The trend toward decentralization could also be seen in the decreasing role played by the central ministries.[47]

This emphasis on self-reliance suggested that such economic concepts as economies of scale, specialization, and comparative costs were largely ignored. However, in 1972-3 the economic policies of Maoism were again increasingly criticized. 1974-5 saw a counter-attack led by the infamous 'Gang of Four'. One of its manifestations was the campaign against Confucius, the real target of which seems to have been Zhou Enlai. In January 1975, at the Fourth National People's Congress, Zhou had given a speech on the Four Modernizations which can be seen as the programme of the moderate group. In 1976 both Zhou and Mao died, and the power balance thereby changed significantly. Without the support of Mao the radicals were isolated and their defeat a question of time.

Soon after 'the decisive victory over the Gang of Four' new economic signals appeared. In late 1976 a front page article in the *People's Daily* declared that China's new leadership intended to step up the nation's purchases of plant technology and capital equipment from abroad:

> Stressing self-reliance does not mean that we advocate a closed-door policy, but we learn from the good experience and advanced science and technology of other countries and absorb them for our own use . . .[48]

The *People's Daily* ridiculed the self-reliance of the 'Gang of Four' by referring to their alleged importation of pornographic films, and emphasized that 'economic and technical exchanges between countries with different social systems are completely normal activities'.

Obviously, the new leadership in Peking did not want to create an impression that a different development strategy was emerging. Therefore a famous speech originally made by Chairman Mao in 1956, entitled 'On the Ten Major Relationships', was used as the major policy document.[49] In this way the emerging policy was legitimized in the name of Mao, although it should be remembered that the situation in 1977 (when the document was first officially published in China) was very different from that in 1956. At that time Mao's mildly critical remarks about the Soviet development strategy were sensational (which is why the document was kept secret); but in the light of later events the same observations sounded rather like praise for the modernization approach.

The wide publication of this document was one of the first important signs of a new trend in economic policies, but for several reasons it took time to stake out the new path. One reason was that the 'gang' had had some sensible points in their criticism against the 'modernizers', although their views had tended to become rather extreme in the course of the power struggle. Further, they had widespread support from those officials who had been recruited during and after the Cultural Revolution.[50] Therefore the preparations for the Eleventh Party Congress, finally held in September 1977, were rather long-drawn out and accompanied by purges at different administrative levels. To the extent that there was a plan for a dramatic change in economic policy (which by now is fairly obvious), there was of course a need for a corresponding change in the power structure.[51] Hua Guofeng was evidently Zhou Enlai's man, but he also owes his present position to Mao's alleged judgment that Hua was the one in the best position to preserve Maoism. He pays lip-service to modernization, but appears to prefer a slower process than Deng Xiaoping.

Deng Xiaoping, on the other hand, has been a major figure in Chinese politics, twice purged (1967 and 1976) but always on his way back again. Deng's economic programme, as for example spelt out in a talk (reported in *China Trade Report)* on the country's economy on 18 August 1975, may be summarized in five points:

1. We must stress the idea of taking agriculture as the foundation. A major task of industry is to promote the modernization of agriculture.
2. We must adopt new technology. To import, we must export a few more things. The first thing to my mind is oil.

3. Step up scientific research work in enterprises. Now some intellectuals have not put to use the skills they have learned.
4. Attaching first priority to quality is a major policy, and this includes variety, specifications, and quality.
5. The key to rules and regulations is the system of responsibility. The present problem is no one takes responsibility.

In the Eleventh Party Congress (September 1977) the modernization drive was firmly set. A 23-man Politburo, committed to economic growth under a banner of order and discipline, was elected, and the policy of the Four Modernizations (agriculture, industry, science and defence) spelt out. The new leadership was particularly concerned about the 'technological gap' created by the voluntarism of the previous decade, and this paved the way for a dramatic change in educational and research policy. The new policy was finally confirmed by the Fifth National People's Congress (January 1978). This was followed by a National Science Conference in March 1978, which saw the wholesale rehabilitation of China's intellectual and scientific community, declared by Deng to be 'a component of the proletariat'.[52] The aim was to have by 1985 an army of 800,000 trained scientific workers in China. Thus, in the new 'Long March' toward modernization, the meritocracy is to play the role of an *avant garde*.

The goal of modernization is no longer disputed but there may of course be many shifts in approach and emphasis as the imbalances and contradictions of the new strategy appear. During 1979 the signs point to a slowing down in speed, suggesting a victory for the gradual rather than the crash approach.

One way to summarize the recent changes in economic policy is to relate them to the six above-mentioned components of the Maoist model. To start with *self-reliance*, there is of course no explicit refutation of this goal. The difference is rather one of interpretation. Today self-reliance is stressed as a reason for importing foreign technology (rather than importing commodities), thus strengthening the productive base of the country. The cultural aspect of dependency is completely ignored. The training of thousands of young intellectuals abroad, thus exposing them to an alternative ideology and way of life, is a case in point. Mao's self-reliance and Deng's do not belong to the same world.

Today *mobilization* is not supposed to be based on moral but on material incentives. Furthermore the trend is toward a hierarchical organization in the factories, putting an end to the chaos of popular participation in decision-making. The revolutionary committees are being abolished at lower levels of administration.

During a period of major policy changes it is not possible to retain a system of *decentralization*. In order to get China on the path of economic growth all sectors of the society must respond to the new signals and make their contribution to the Long March Toward Modernization.

As an article in *Beijing Review* (15 June 1979) made clear:

> . . . modernization requires a political environment of stability and unity, stable and good social order, strict regulations and rules and labour discipline and centralized, unified command.

As far as the three contradictions (*urban-rural, agricultural-industrial* and *intellectual-manual*) are concerned, the dialectic process has obviously taken a turn in a non-Maoist direction. Deurbanization was a major component in the Maoist model and it is likely that during the sixties China succeeded in at least keeping its cities from growing faster than the countryside.[53] This policy has been terminated and many town-dwellers, earlier dispersed in the countryside, are now returning to the urban world with aggressive demands for re-employment and rehabilitation. The policy of modernization is obviously not compatible with continued deurbanization. Rural industries are still emphasized but rather in terms of their overall functions in the industrial system than as an asset in themselves, or as a means to achieve self-sufficiency at the commune level.

The new relationship between industry and agriculture is still unclear. The initial emphasis on industry (particularly steel) in the modernization programme was modified during the 1979 'slowdown' and the importance of the agricultural sector was stressed. Within this sector a significant shift from grain production to the production of cash-crops took place. Regional specialization according to the principle of comparative advantage and productivity-raising measures were also stressed. All this points to a more conventional pattern of economic development. Like India, China will also have its Green Revolution.

Finally, the intellectual-manual contradiction illustrates most dramatically and explicitly the shift in development strategy. It is necessary to see the educational system as an integral part of the total development strategy. Whereas the Maoist model necessitated a very specific kind of training, stressing political consciousness and a command of very diverse skills, such an education will only turn out to be a handicap if the process of economic development is based on imported technology. The latter pattern of development implies specialization and professionalism. Since it takes time to develop new skills it is in this field that the measures to make up for past 'mistakes' are bound to be most dramatic, not to say desperate.

These changes are not only incompatible with Maoism but they may, at least in some of their consequences, be difficult to reconcile with any variety of socialism. A reported trend toward the spontaneous disbanding of the production teams is a case in point. Such a process would undermine the most cherished innovation of the Maoist strategy, the people's communes. The policy of agricultural specialization, and the downgrading of grain production in order to boost production of cash-crops for export, will have similar consequences. Occasionally the communes are even said to be obstacles to the modernization of agriculture.[54]

It is at present not possible to assess the importance and significance of these and similar signs. Perhaps we are too alarmist? Or could it be that socialism is at stake in China? Scholars who have made important contributions to our understanding of China disagree intensely on this issue today.

In a statement which has attracted much attention, Charles Bettelheim, a great admirer of the Maoist model, made the allegation that the apparent fidelity to Mao Zedong's policies has been a 'smoke-screen designed to conceal a quite different line.' In his view 'a revisionist line is presently triumphing.'[55] Many China watchers found this judgment a little rash and instead suggested that the new measures were designed to cope with new problems or correct former mistakes. A distinction was made between 'strict' discipline and 'exploitable' discipline, the former associated with industrial socialism the latter with industrial capitalism.[56] The critics also emphasized the atmosphere of relief and freedom prevailing in China after the fall of the Gang of Four.[57] Another important point raised by the critics was the error of inferring social practice from the political line: 'Today the line has shifted, but the experience of the earlier period gives us every reason to believe - contrary to Bettelheim - that practice has not changed nearly so much.'[58] Thus there are two lines of 'defence': recent changes are necessary; or they will have only a limited impact. If they are 'necessary', they are so only to the extent that a modernization on the lines of Western industrialism is the ultimate goal for all societies. As to the impact, it may be true that a policy that goes against the forces of modernization will face great difficulties and much sabotage. But present policies are with the current. They are bound to have an impact.

Concluding Observations

As the preceding overview has shown, the development strategies of India and China have followed winding roads. In India the dialectics of development seem to be at work primarily at the policy level with limited impact on economic structure. Certainly the emphasis on public investment in the heavy industry sector in Nehru's days and the new agricultural strategy from the mid-sixties and onwards deeply influenced the pattern of economic and social development, but on balance development planning has more and more turned out to be a somewhat ritual exercise in India.

In China the shifts in emphasis, as far as economic policies are concerned, have had more important structural implications, simply because of the nature of the political system. It must, however, be emphasized that our general lack of knowledge about what really goes on at the local level in different parts of China makes it too easy to exaggerate the real effects of different signals from Peking.

Another observation is that the movements of the pendulum have gone in opposite directions in India and China. Thus when one country stressed agriculture the other stressed heavy industry; when one followed outward oriented policies, the strategy of the other was inward oriented; and so on. On the whole the dialectics have worked in a more constructive way in China. The issue of self-reliance is a case in point.

There are few countries in the Third World having a real option with regard to 'self-reliance' versus 'modernization', as here defined. India and China are among those few, and it is also in these two countries that the option has been a major political issue for an extended period. Both of them have large natural resources, huge domestic markets, and great cultural traditions to draw upon. Nevertheless the traditions of Westernization are also long and important, since modernizers in both countries have made use of Western ideologies such as Liberalism and Marxism to combat indigenous conservative traditions (Brahminism, Confucianism). The temptations of the Western-inspired policy of modernization have therefore been substantial, in spite of the well-known dangers of dependence. In discussing the power structures we concluded that the forces of modernization and self-reliance were not equally strong. Whereas the 'modernization paradigm' receives authorization from the Western experience or unrivalled technological change and economic growth, the untrodden path of populism, self-reliance and 'alternative' patterns of development can rely on little else than the charismatic power of inspired leadership. History indicates that this is insufficient. Furthermore, both India and China have recently been deprived of the active incarnations of such traditions: Jayaprakash Narayan, who operated within a revitalized Gandhian tradition, and Mao Zedong, whose message is becoming more and more distorted as the new leaders consolidate their position. Significantly the utopian visions of both leaders were compromised and frustrated through the opportunism of political 'followers'.

For the present the economic policy of China is outward-looking, whereas the economic policy of India's Janata government until its fall in July-August 1979 was oriented toward self-reliance, decentralization, employment creation, small-scale industries, rural development etc. It should be noted that these were also the components of the Maoist development strategy. In India little was achieved in this direction, however, since the government spent its energy in a desperate internal power struggle. Our analysis of the underlying power structure of the Janata 'party' showed why this had to be so. As late as November 1979, it was impossible to foresee what political combinations might emerge from the unprecedented political disintegration in India.

However, the era of dominant party rule seems to be over and a new era of short-lived coalition governments is in sight. If this scenario comes true, the relevance of what goes on in the political superstructure for the actual process of economic and social change may be even less than it is today. Thus the spontaneous process of 'development' will continue unabated, a process that will benefit the forces of 'modernization' in 'growth areas' rather than those of 'self-reliance' - or, for that matter, an optimal mixture between the two.

In contrast with the situation in China in the nineteenth century (the Self-strengthening Movement) and India today, China can carry out its present modernization programme from a position of relative strength. There is no question of accepting external inputs to make up for a *general* lack of internal inputs. Rather the point seems to be a *selective use* of external inputs in order to eliminate important bottlenecks in development, in exchange for a resource which China more than most other developing countries can spare: namely oil. If the output of oil turns out to be less than so far has been expected, which seems likely, there will be an increased export of cash crops and products from light industry. At present this strategy seems to be preferred over the more dangerous policy of budget deficits and borrowing abroad.

Instead of being an exporter of ideology, China will be an exporter of oil. The change is significant. Today China is in the process of re-evaluating and demythologizing the accomplishments of Mao Zedong in an attempt to change the Chinese collective consciousness. It would be undialectic to say that this attempt will succeed in the long run, but at present it is rather obvious that anti-Maoist policies have substantial popular backing. It should also be obvious that these policies are bound to create new imbalances and contradictions. Those (including the present author), who persist in admiring the Maoist strategy as a solution to China's problems and as a possible model for developing countries, set their hope on a new turn of the pendulum. A few pessimistic reflections are, however, called for. A movement from 'chaos' to 'order', from amateurism to professionalism and from mass participation to bureaucracy is obviously easier to bring about than the reverse. The Great Leap and the Cultural Revolution were extraordinary incidents made possible by extraordinary circumstances, among which we hold the charismatic power of Mao Zedong to be the most important. What we now are witnessing is the 'routinization' of charisma, a change from revolutionary charisma 'in the process of originating' to the system-maintaining 'charisma of office'. However, the Chinese revolution belongs to the Chinese.

NOTES

1. For a discussion of the Western Model of Development in the context of intellectual dependency and liberation, see B. Hettne, *Current Issues in Development Theory,* SAREC Report R5:1978.
2. In Latin America 'populismo' signifies a nationalist development strategy, based on a class alliance of the national bourgeoisie and the labour class (or rather the trade unions), and aiming at industrialisation. In Asia (and Africa) populism (much like the Russian Narodniks) is oriented towards rural development, the local community and the indigenous culture.
3. Very much the same challenge was created within Europe through the British industrial lead; some countries were able to take up the challenge and some were not. The various responses to this challenge, which gave a specific pattern to European industrialization, are dealt with in A. Gerschenkron, *Economic Backwardness in Historical Perspective,* Cambridge, Mass. 1962.
4. On this, see Mary C. Wright, *The Last Stand of Chinese Conservatism: The Tung-chih Restoration,* Stanford 1957.
5. Benjamin Schwartz, *In Search of Wealth and Power. Yen Fu and the West,* Cambridge, Mass. 1964.
6. Chow Tse-tsung, *The May Fourth Movement. Intellectual Revolution in Modern China,* Stanford 1967, p. 359.
7. Dadabhai Naoroji, *Poverty and un-British Rule in India,* London 1901.
8. Ssu-yu Teng and John K. Fairbank, *China's Response to the West. A Documentary Survey 1839-1923,* New York 1975, p. 19.
9. B. Hettne, 'The Vitality of Gandhian Tradition', *Journal of Peace Research,* No. 3, Vol. Xiii, 1976, p. 20.
10. B. Hettne and G. Tamm, 'The Development Strategy of Gandhian Economics', *Journal of the Indian Anthropological Society,* Vol. 6 No. 1, 1971, p. 54.
11. Jerome Chén points out that Mao's impatience with bureaucracy stems more from Chinese than Western sources. J. Chén, *Mao and the Chinese Revolution* Oxford, 1965. See also Stuart Schram, *The Political Thought of Mao Tse-tung* New York 1969.
12. Maurice Meisner, *Li Ta-chao and the Origins of Chinese Marxism* New York, 1970.
13. Maurice Meisner, 'Leninism and Maoism. Some Populist Perspectives on Marxism-Leninism in China' *The China Quarterly,* January-March 1971.

14. Ibid., p. 16.
15. For a comparison between Gandhism and Russian Populism see Hettne, 1976, op. cit.
16. Hettne and Tamm,1971, op. cit.
17. J. Bandyopadhyaya, *Mao Tse-tung and Gandhi* New Delhi, 1973; R. Vaitheswaran, 'Gandhi and Mao: A Comparison in Terms of Relevance for the Politics of National Liberation and Reconstruction' *China Report,* Vol. XII, Nos. 5-6, 1976; J.D. Sethi, 'Mao and Gandhi: Convergence and Divergence - A Preliminary Note' *China Report,* Vol. XIII, No. 1, 1977.
18. J. Galtung, *The Politics of Self-Reliance,* Chair in Conflict and Peace Research, University of Oslo.
19. Sethi, op. cit., p. 26.
20. Vaitheswaran, op. cit., p. 35.
21. Sethi, op. cit., p. 27.
22. Vaitheswaran, op. cit., p. 35.
23. Sethi, op. cit., p. 28.
24. For a more historical approach see *B. Hettne, Utvecklingsstrategier i Kina och Indien* (Development Strategies in China and India) Lund 1971, 1979.
25. After the split in January 1978 there were four 'congress parties', two inside and two outside Janata. Many leading figures in the Janata government were former congressmen.
26. See the discussion in P. Worsley, *The Trumpet Shall Sound* New York 1968.
27. See an article by Dev Murarka in the Danish newspaper *Information,* 29 April 1977.
28. S.P. Cohen, *The Indian Army,* Berkeley, California, 1971, p. 187.
29. G.H. Corr, *The Chinese Red Army,* Osprey 1974. For a review of books dealing with the relations between the People's Liberation Army and Chinese Society at large, see Lynn T. White III, 'The Liberation Army and the Chinese People', *Armed Forces and Society,* Vol. 1, No. 3, May 1975.
30. This oscillation approach has been critically discussed by A.J. Nathan ('Policy Oscillations in the People's Republic of China: A Critique', *China Quarterly,* 68, 1976), who notes that there are authors who use the policy oscillations model as 'a kind of shorthand, convenient for summarizing the content of policy and its changes', but that there also are authors who want to explain policy oscillations. Among these he distinguishes three categories: (1) those who see the oscillations as cyclical movements between Communist utopia and the

realities of backwardness; (2) those who see them as conscious applications of Mao's conception of dialectics and (3) those who see them primarily in terms of changing power relations. The previous discussion of power structures would probably qualify the present author for membership in the third category. Nathan's criticism may be summarized thus: (1) It is not possible to place all policy options under two headings; (2) it is not possible to establish a necessary link between different sets of 'right' and 'left' policies; (3) the 'two line' struggle is so forcefully projected by the Chinese media that outside analysts are led astray. To this our reply would be that it is quite enough if certain fundamental perspectives (for example what we here have called 'self-reliance' versus 'modernization') could be contrasted. It is not necessary to explain *every* programme in terms of policy oscillations. Secondly, we would agree that it is very difficult to establish a close connection between internal oscillations and Chinese foreign policy. Possibly they must be analysed within different frameworks. As for internal linkages there are certainly causative relations between some if not all areas. Thirdly, there has obviously been a sometimes very bitter and bloody political struggle in China which simply does not fit with a less dramatic 'trial and error' model of development.

31. For the concept of 'reconciliation system' and 'mobilization system', see D.E. Apter, *The Politics of Modernization,* Chicago 1965.
32. We are here mainly relying on Hettne (1979).
33. JP said: 'I'm afraid my Sarvodaya friends may not like this, but I have come to feel that our approach was wrong . . . The struggle approach is the right one. Gandhiji had seen this when he developed struggle along with constructive work'. *Indian Express,* 13 July 1974.
34. Personal observations from Karnataka.
35. 'Janata Party Manifesto', published in *Commerce,* 19 February 1977.
36. *Times of India,* 29 March 1977, p. 7.
37. *Times of India,* 4 May 1978.
38. *Economic and Political Weekly,* 28 January and 4 February 1978.
39. *Economic and Political Weekly,* 14 July 1979.
40. G. Ostergaard, *JP's Total Revolution: in Retrospect and Prospect,* (mimeo 1979).

41. Ibid., p. 16.
42. *Far Eastern Economic Review* 31 August 1979.
43. *Economic and Political Weekly* 18 August 1979.
44. Lowell Dittmer *Liu Shao-chi and the Chinese Cultural Revolution*, Berkeley, California, 1974, especially chapter 6.
45. Ibid., p. 180.
46. J. Sigurdson, 'Rural Industry - A Traveller's View' *China Quarterly*, No. 50, 1972.
47. A. Donnithorne, 'China's Cellular Economy: Some Economic Trends Since the Cultural Revolution' *China Quarterly*, No. 52, 1972.
48. Quoted from *China Trade Report*, January 1977, p. 8.
49. *Far Eastern Economic Review*, 4 February 1977, pp. 10-12.
50. Leo Goodstadt, 'Poverty and Power in Peking Politics' *Economic and Political Weekly*, 3 December 1977.
51. In January 1977 a wall poster which appeared in Guangzhou spelt out a modernization strategy in nine points, recommending among other things increased levels of consumption, material incentives, workers' control, and rapid introduction of foreign technology. Significantly, the poster was anonymous. (*China Trade Report*, March 1977, p. 9).
52. *Asiaweek*, 14 April 1978.
53. C.P. Cell, 'Deurbanization in China. The Urban-Rural Contradiction', *Bulletin of Concerned Asian Scholars*, Vol. 11, No. 1.
54. *China Trade Report*, May 1979.
55. Letter of resignation to the Franco-Chinese Friendship Association. The letter was published in *Monthly Review*, July-August 1978, and followed by an interesting debate between Bettelheim and Neil Burton. Later a special issue of *Monthly Review* (May 1979) was devoted to the same debate with contributions from, among others, Jerome Ch'en and Joan Robinson.
56. R.S. Leiken in *Monthly Review*, May 1979, p. 37.
57. According to Joan Robinson the present Chinese leadership 'has embarked upon a hitherto unprecedented course of combining an ambitious plan for accumulation and growth with open discussion and freedom of thought.' Ibid., p. 56.
58. A. MacEwan, ibid., p. 45.

SCIENCE AND TECHNOLOGY IN DEVELOPMENT: POLICY OPTIONS FOR INDIA AND CHINA

Aqueil Ahmad

Many, perhaps most, China experts would advise against comparing China with other countries *vis-à-vis* developmental efforts and corresponding gains and losses. The Chinese experience, or the 'Chinese Model' as it is often called, is supposed to be entirely unique on account of China's peculiar history, her vast size, enormous human and other resources, and above all, the radical social, political and economic organization of the post-revolutionary period.[1] While a sizeable body of literature on various aspects of Chinese society has been gradually accumulated in the English language over the past twenty-five years, comparative analysis between China and other societies has not been very popular. The country which is, however, most often compared with China is India - despite well-recognized hazards. The most common theme in India-China comparisons undertaken by Indian and foreign scholars alike is economic performance of the two societies during the past three decades or so, despite the fact that the Chinese data is often unavailable or is considered not completely reliable. The latest among comprehensive quantitative analyses of this type are the studies of Sarwar Lateef and Subramanian Swamy.[2]

Comparative studies on India-China science and technology systems are still rarer. To the best of my knowledge, nothing of this sort has ever been undertaken either in India or China. The very first study, of a purely statistical nature, was conducted by Rushikesh Maru in 1969 at the Research Policy Program at Lund University.[3] It was not followed by similar efforts until several years later when Jon Sigurdson produced a series of comparative studies in Sweden on India-China rural and urban technology systems and their role in economic development. He has continued to work in this direction ever since.[4] A very interesting, strictly policy-oriented comparative study of science in India and China was completed at the Department of Political Science, University of Melbourne in 1976.[5]

The Purpose and Scope of the Present Analysis

As mentioned earlier, the complexity of issues involved as well as paucity of relevant information and comparison criteria do not permit a one-to-one comparison between India and China on all counts. However, if the experience of the two societies is to be meaningful to each other at any level and to any degree, comparisons of some kind must be made - though not necessarily to prove either system right or wrong or better or worse. India and China can be compared at the following two levels:

1. at the highest level of generality, as total entities with broad similarities and differences in their peculiar historical and cultural circumstances, purposes, strategies and their own goal-related successes and failures in one or more aspects of national life, and

2. as detailed comparative analyses of certain carefully selected identical areas, problems, conditions, issues, etc. and the way these may have been 'solved' or handled in the two societies.

The comparative overview of science and technology policy issues in relation to national goals and developmental objectives in India and China intended here falls more in line with the first approach. Since science and technology policy options cannot and must not be examined in a socio-economic vacuum, a considerable portion of this essay will be devoted toward construction of a larger framework of conditions and circumstances in the two societies within which science and technology policy issues would become more meaningful. Towards the end, a brief commentary will be presented on the relevance of India and China for each other, for other Third World countries and for the rest of the world.

The Starting Point: Comparative Socio-Political and Economic Circumstances

Despite vast differences in their history and social structure, economic and physical conditions in India and China were remarkably similar in many ways at the time they embarked on the path of planned socio-economic development some thirty years ago. The common factors of a large land mass, huge populations, low agricultural productivity, urban-rural schism, age-old exploitation of the peasantry by local elites, landowners and colonial rulers, predominantly agrarian nature of socio-economic organization, proverbial poverty of the people, etc. are universally known.

Given their peculiar recent political histories - Marxist revolutionary struggle versus Gandhian non-violent change - the two countries embarked on a course of economic development and social change with identical goals, but adopted radically different approaches to reach these goals. Both were deficient in physical resources and technological skills required for the mammoth tasks ahead. The difficulties for the new leadership in the two countries were further compounded by the holocaust of partition and resettlement of millions of refugees in the Indian subcontinent and, in the case of China, by the continuing struggle between the revolutionary forces and the diehard bourgeoise elements within Chinese society, as well as the strong anti-Chinese feelings and withdrawal of support for China in the international community led by the United States under the shadow of the Dulles doctrine and supported by such industrially advanced nations as Japan, an arch-enemy of China in the Orient.

The foundation of the new China was laid by Mao Zedong according to Marxist ideology suitably adapted to the Chinese conditions. How a non-Chinese European system of thought could be so successfully implanted in China within a short period of time bespeaks of the genius of Mao and his profound understanding of Marxism as well as of Chinese society, its people and their needs. Today China can be cited as the purest type of Marxist society approximated anywhere in the world. Is there, or was there, something intrinsically suitable in the Chinese culture and philosophy and social structure to make China so hospitable and amenable to Marxism? The answer to this riddle may help speculate about the future of Chinese Marxism as it moves along and evolves through fast-changing needs and complexities of that society and its rapidly advancing economic and technological frontiers internally as well as globally. For the present, Maoism or the Chinese Marxism, has served China well. The strategy of totalitarian management of society through a highly centralized party leadership, mass appeal and military backing, coupled with decentralized decision-making for decentralized agricultural and industrial production, has generated the necessary motivation and commitment of the millions so essential for the urgent task of increasing productive capacity of society for achieving well-established national goals of enormous proportions. Chinese science and technology policy and its implementation can only be seen and understood within this grand strategy, this overall national/societal framework whereby science and technology have a definite practical social purpose as opposed to the notion of 'science for science's sake'.

Under the influence of Gandhi and Nehru, India opted for a democratic form of government and society. Modern Indian institutions have evolved in the liberal-democratic traditions implanted in the country through British-style education and thinking of the Indian elite. It is often said that democracy is not suitable for India, and that the story of Indian development would have been entirely different had India gone the totalitarian way. But as a matter of fact, given the diversity of Indian life and culture, her philosophical and intellectual traditions, her former English rulers, her Gandhi and Nehru, democracy seems to be ideally suitable for the Indian way of life. The issue for India today is not totalitarianism versus democracy but how to make the democratic government work more efficiently and honestly for the good of the people it supposedly represents. Indian developmental effort and the role of science and technology in development will have to be assessed within this overall social and political framework. If the Chinese experience, or for that matter the experience of any other country, is to be useful to India, it will have to be applicable within that framework. In any case, the overall developmental goals for India and China were remarkably similar: to remove the poverty of the masses, feed, educate, clothe and house them properly, and to improve the overall productive capacities by resource mobilization and modernization of systems of production in industry and agriculture. The means adopted for reaching these goals were, however, vastly different in India and China.

Alternative Strategies of Socio-Economic Development and Resource Mobilization

Under a totalitarian, highly centralized national-level planning and leadership, China emphasized, and largely succeeded, in generating mass support and involvement at various levels of developmental decision-making and action. The best examples of this collective approach are large-scale construction projects and by now world-renowned people's communes, production brigades and teams where responsibility and accountability for setting targets and other practical matters and their consequences have been delegated to the people themselves. The worst examples may be a strirctly controlled information system, over-politicalization of issues and personalities, and coercive tactics for ideological and stylistic conformity witnessed in different periods, most notably during the Great Proletarian Cultural Revolution of the late sixties and, once again as a reminder, in the denigration of the so-called Gang of Four.

The policy as well as action for development in democratic India - where the social structure remained by and large unchanged and maintained its continuity with the past - emanated from the urban-based educated elites, experts, leaders and bureaucrats. Development in India was seen as something 'to be done for the people' but not necessarily by the people themselves. The best examples of this approach are the Community Development Programme and the Panchayat Raj system started in the early fifties, rapid industrialization and modernization of certain selected sectors, and building up of impressive knowledge and expertise bases on international standards. The worst examples may be the elitist and fashionable nature of so many developmental programmes (i.e. nuclear and space research and development and luxury consumer-goods industry), bureaucratization of change, lack of coordination and implementation of otherwise sound schemes, and above all, the remaining poverty of the masses and the remaining - or even widening - gap between the rich and the poor. It is inconceivable to think of the Chinese level of inequality reduction in India given her elitist orientations and the prevalent structure of vested interests, which are reflected in almost all the institutions of society, including those for science and technology research and development.[6]

India and China both had decided early to achieve scientific and technological self-reliance and to use science and technology in large doses for developmental purposes in order to achieve economic self-reliance. Both countries started out by receiving techno-economic aid from outside while at the same time building their own capabilities. For China, it was the Soviet Union all the way until that relationship was broken during the Kruschev era and China began to learn to depend more and more on its own indigenous resources. Many China experts claim the break with Russia was the best thing that could have happened for China. It may, however, be noted with interest that at least some seasoned China watchers are beginning to see things differently. The recent policy goal of reaching 'the advanced levels of the 1970s in a number of important branches of science and technology' during the next eight years could mean an admission of inferior status for China as well as admission of superiority of the advanced technological systems in the Western world. If that be so, one can argue whether the Russian pull-out in the early sixties was detrimental to China's long term scientific and technology capability-building efforts. In any case, the very notion of 'catching up' would imply that after one and half decades of self-imposed technological isolation, the new Chinese leadership may once again be thinking of looking Westward.

India, under the stance of political neutrality, received assistance from various sources both in the capitalist as well as the communist camps, although the bulk of external input was provided by the former in the form of technical and economic aid or international collaboration in manufacturing industries. Unlike China, India's technological dependence on advanced industrial societies has not only continued but increased in certain areas as Indian industries have tried in ever larger measures to innovate, increase profits and compete in the newly-emerging international industrial and economic order. Large-scale, modern industrial sector in India has shopped around for technological innovations in the international technological marketplace while the efforts of the local research and development establishments have remained largely unutilized. Multinational corporations and other foreign firms have a strong foothold on industrial production in India through equity shares ranging anywhere between 40 to 100 per cent in certain cases, but more importantly through the control of technology innovations and technology transfer arrangements. At present, about 180 multinational corporations have 740 affiliate organizations, 538 branches, and 202 subsidiaries in India. British, American, Swiss, German, Swedish and Japanese firms predominate. Their aggregate capital investment amounts to 30,000 million rupees (approximately US$ 3750 million). Estimates of proportionate capital investment of foreign firms in India vary between 5 to 15 per cent of the total investment, the remainder coming from exclusively Indian sources. The Foreign Exchange

Regulations Act of 1973 stipulates dilution of equity shares from 100 per cent to 74 per cent for companies operating in priority, export oriented and/or sophisticated technology areas. For others it is to be 40 per cent maximum. In all types of foreign collaboration for industrial production, whether through the agency of multinational corporations or other foreign firms which are not multinational corporations the industrial policy has laid heavy emphasis on technology transfer, research and development capability-building, export markets, and general Indianization of the firm. Royalty payments for the import of technology have been discouraged. But multinationals have often failed to comply with the requirements.[7] For example, the proportion of research and development personnel to the total employed was only 1.4 per cent in purely technical collaboration companies, roughly 2.5 per cent in the subsidiaries of the multinational corporations.[8] It is generally recognized that the export performance of Indian firms with foreign collaboration of the subsidiaries of multinationals has not been significantly better than those without such affiliations, although a substantial portion of such agreements are entered into with the plea that they would help augment exports of the indigenously produced goods. One of the prime reasons for this deficiency is the multinationals' desire to capture local markets rather than increasing the exports share of the host firms in world markets which are largely controlled by the multinationals' global marketing operations and over which the local partners have no control or claim.

There are no multinational corporations in China. After the Chinese paid off the Russian debt of the fifties, there has been no foreign capital investment either. It is however, well known that the Chinese benefitted greatly from the Russian technology and capital in their initial developmental thrust - especially in building up heavy industrial capabilities - for nearly a decade before that relationship with Russia was broken off. An authentic quantitative estimate of Russian capital and technology in China, or its real impact on the Chinese economic development, would be next to impossible to make. One of the most astute China observers, the late Edgar Snow, contended many years ago that the magnitude and impact of such technical and economic assistance from Russia was quite substantial for China initially, but that later on the Chinese mastered the technology themselves to the extent that they could adapt, innovate and in certain areas and up to a certain point develop on their own.[9] That capability may have reached a saturation point several years ago and Chinese technological advancement may have suffered a serious setback due to the absence of a steady supplier of advanced technology. There are clear indications that the Chinese may soon be shopping around in the international technological marketplace for buying or borrowing sophisticated technologies for at least some of their special needs in order to reach or maintain a commanding position among the community of nations. To a degree, the current interest of industrially advanced nations of Europe and America in the Chinese development - or for that matter the development of the Third World in general - can be attributed to purely business interests looking ahead to vastly expanding unlimited markets for Western technology.

The Russian technical aid to China came without any political or economic strings. It was aid from one communist country to another sister communist country, an aid designed for the building up of world communism. This fact of tremendous significance is generally ignored in assessing the role of international technology in Third World development, particularly for large countries like India that had to import foreign technology sometimes under adverse terms and conditions in order to produce basic materials and machinery needed for further developmental tasks. The Russian manner and mode of technical and financial assistance to China during the fifties may indicate that unfettered international technology and capital can be powerful and effective tools of modernization and development of needy nations, provided of course that the latter are willing and eager to learn from others so that they could ultimately become strong enough to stand on their own feet.

In the Indian case, the role of international technology and capital transferred through the agency of multinational corporations and other types of foreign collaboration in building up industrial technology capabilities, as well as their impact on the economy in general, is highly debatable. The highest degree of foreign association involving direct foreign involvement and technology has been witnessed in the capital-intensive consumer industries owned and controlled by the private sector, serving the needs or fancies of the urban economic elite. However, the situation may not be viewed entirely negatively, for different analysts sometimes present diametrically opposed pictures depending upon the intended frame of reference. For instance, while K.K. Subrahmanian and Peter Neersø present a highly negative view of technology transfer through foreign collaboration and its impact on the Indian economy and indigenous technology capability building process,[10] K. Jayaraman in a recent analysis tries to show how foreign association in Indian industries has helped build technological self-reliance, augmented exports, streamlined production and management, encouraged reinvestment of profits in India, saved foreign exchange, etc. His conclusion: International investment and multinationals, under proper regulations and controls, are inevitable for developing nations.[11] Such controls are to be provided within the framework of the Foreign Exchange Regulation Act of 1973 and, more importantly, by the Janata government's new industrial policy which proposes a much tougher attitude toward the behaviour of foreign firms without altering the basic position of continuing to allow foreign capital and technical inputs in India.

The problems of rural agro-industrial technologies and development are of a different nature in both India and China. In both cases there has been little or no direct application of foreign capital and technology. The Chinese and Indian accomplishments and setbacks here are to be explained more by sociological than technological factors. Unfortunately most of the comparative sociological analyses of Indian-Chinese development provided by American scholars are superficial and quite often of dubious quality.[12] Suffice it to say for the purposes of this analysis that the Chinese success in the countryside is due mainly to the radical form of social organization (abolition of private property and political power, land reforms followed by establishment of peoples' communes, propaganda machinery, control over private consumption, wages and incentives, the ideological push, etc.) introduced, even enforced, by the central leadership. It would not be sufficient to say that totalitarianism, Mao Zedong, or the Chinese temperament alone were the prime movers in China while democracy and lack of 'social development' prevented a large-scale transformation of Indian society, especially in the rural areas. Other totalitarian systems with sufficiently strong leadership have failed to accomplish similar (to China) goals elsewhere. A thorough sociological analysis must by necessity point out the processes through which such radical reforms have been introduced in a relatively short period of time, and whether and how under different circumstances similar reforms could be attempted in other developing societies at different stages of development. Needless to say, nothing on that grand scale has so far happened in the Indian countryside. The Community Development Programme which started in great earnest in the early fifties was soon bureaucratized and became part of the centuries-old district administration where the District Magistrate is still the supreme commander. The seriousness of the Indian situation has not quite sunk into the consciousness or behaviour of the so-called district level developmental administration which remains to this day the central mechanism for rural development. Despite these constraints, agricultural production in India has doubled during the past 20 years. India today has nearly 20 million tons of foodgrains surplus although many people still go hungry at least some of the time. The role of new agricultural technology in raising farm productivity has been substantial, although the rural socio-economic structure has allowed the greatest benefits to accrue to large, already prosperous farmers. Much of this technology has come out of indigenous efforts at adaptation or innovation in which the Indian Council of Agriculture Research has played a signal role. Why indigenous agricultural research has been so much more instrumental in agricultural development in India than industrial research in industrial development is a subject of considerable interest and importance, but that is a story I have told elsewhere. [13] Besides the failure at land reforms, one of the main drawbacks of Indian rural development strategy, unlike that of China, has been lack of integration of agricultural with corresponding industrial development, lack of integration of agricultural technology with rural industrial technology. Jon Sigurdson has alluded to this circumstance in some of his earlier quoted writings. What is of further interest is the claim by the current leadership in New Delhi that the new industrial policy would correct some of these imbalances. A careful analysis of the so-called new policy seems to belie this claim, and hope. The policy lays considerable emphasis on small industry, while ignoring the rural 'mini industries'. Small industry is structurally linked with the capitalist class in India in that it is financed and controlled by the urban-based petty bourgeoise, thrives on subcontracts of large industrial and trading houses, and consequently ends up serving the urban consumer class. On an overall basis, the new policy fails to pinpoint how the proposed new industrial order is supposed to serve the interests of the common man, or how the gap between agricultural and industrial development is going to be removed in order to evolve a more comprehensive development plan for rural transformation.[14]

Science and Technology as Resources

Ninety per cent of science and technology research in India is financed and controlled by the government, the remainder by private industry. India has evolved a complex and multilayered system of policy, planning and management of scientific and technological research whereby national goals are supposed to find expression in research and development priorities and objectives at the operating institutional levels. Total expenditure on science and technology today amounts to Rs. 1535.3 million annually (US$ 192 million approximately), which is equivalent to 0.5 per cent of India's gross national product. I believe China is already spending around 1 per cent of her gross national product for science and technology research and development - a percentage which India too would soon be reaching for her own research and development expenditures, most probably by the end of this decade. Although small in relation to their needs, both countries have large stocks of qualified scientific and technical manpower in absolute terms, being alternatively either third or fourth in international comparisons. Relative to their limited resources, these are large investments for India and China; although relative to their needs they may be quite small.

Much of the Indian research and development work is carried out in autonomous research institutes/laboratories from which the transfer of technology to the users has posed nagging problems for research planners and managers. In China a great deal of technological adaptation and innovation has been done on the shop floor through the so-called 'worker innovation' system or agricultural experimentation carried out by decentralized research stations at the commune level. It is, however, often claimed that while this 'shop floor' variety of research, adaptation and designing may have given China an immediate, short-run edge over technological dependence and production slacks, it has not resulted in long-term scientific and technological capability so essential for building stable foundations for techno-economic self-reliance.[15] India's experience by implication alone, would be just the reverse of that. Under the influence of liberal scientific traditions, Indian scientists, educators and intellectuals in general have jealously guarded against any encroachments on intellectual and professional freedom. Their work has not always been undertaken with well-defined social objectives similar to their Chinese counteparts' strictly practical orientation under the ideology of a socialist science. The Chinese have done bold experimentation with their science and technology systems in order to make them socially responsible and practical. The notion of 'mass science' has no parallel anywhere. To make the idea of mass science a programme of action, scientists have been thoroughly indoctrinated into Marxist ideology. They have been encouraged, or even forced, to go out into the fields and work with farmers and peasants while the latter have been invited to participate in educational and research programmes in communes, production brigades and teams.[16] Science in China is done for a collective purpose whereas in India it is undertaken as private activity or private entrepreneurship, although this orientation is gradually weakening in national research institutes and laboratories. It is quite conceivable that while China has succeeded in bridging the gap between science and society, theory and practice, research and application, scientists and peasants,[17] India may have acquired potential scientific and technological skills and infrastructure which, under the right kind of political support, administrative approach and institutional reforms, may ultimately give India an edge over China in the long run. It is often claimed that one of the reasons for lack of application of science to practical problems in India is that most scientific research is essentially of a basic character. My own assessment is that this is not quite the case. If anything, Indian science, like its Chinese counterpart, is lacking in basic research. It would not take much to arrive at this conclusion if one carefully examines the lists of projects undertaken in national research institutes and laboratories, although the situation in university research departments may be different. Project proposals in India today are strictly evaluated on their practical relevance, unless of course the project specially calls for basic research. Sadly enough though, the practical relevance quite often remains in theory only, for science in India is not strictly related to production, as in China.[18] To put it more directly, there are no takers of Indian research, especially industrial research. Prospective users of industrial research either do not know of the Indian expertise or deliberately ignore it in favour of foreign technology for reasons of expediency, risk coverage, prestige and other economic and cultural reasons. While the Chinese science has evolved out of and for production needs, the Indian science has emerged as a grand abstraction out of the 'limbo effect' of national development where people, systems, activities could easily acquire a ghostlike characteristic of being there without their feet on the ground.[19]

The Indian scientist, unlike his Chinese counterpart, especially after the Cultural Revolution, has not been politicized and ideologically motivated to work toward national goals. He considers himself a scientist first and foremost, above and beyond politics or even society. Unlike the forces unleashed for this purpose during the Cultural Revolution, the process of new consciousness generation among Indian scientists has taken on a slow and unsteady course through specialized education and a mild public debate often marred by laying blame for the 'failure' of Indian science on this or that type of science or scientists. In the current thinking of the new Chinese leadership, politicalization of science and scientists in the wake of the Cultural Revolution may have gone too far and needs moderation in order for science to be professionalized once again according to international standards.[20] There must be a happy medium somewhere for developing societies like India and China to combine professional excellence with social responsibility - in science as well as in other aspects of national life.

Indian and Chinese science and technology policies have continuously emphasized the goal of self-reliance. But in China the objective has also been to not only catch up with technologically advanced societies but at some stage to beat them and surpass them. Mao voiced such concerns on a number of occasions in unmistakable terms. In keeping with the Chinese style, this may have been a political gimmick, but one can seriously ask the question: Why not? After all, the Chinese (or for that matter, the Indians and Arabs) had at one time achieved great heights of scientific and technological advancement while Europe languished in the Dark Ages. And one set of societies cannot and must not be expected to maintain techno-economic superiority forever. Have not Europe and America held it for long enough since the Industrial Revolution?

Reflecting the Maoist idea of 'perpetual revolution', the Chinese science and technology systems have been subjected to a number of deliberately introduced upheavals during the last three decades. Three distinct phases of Chinese science can be identified:

1. Immediate post-revolutionary period - marked by vestiges of colonial science, social isolation, and uncertainty at the policy level regarding the character of science in revolutionary China.

2. Marxist science: Science in the aftermath of the Cultural Revolution - marked by radicalization of scientists and scientific institutions, nationalist-socialist overtones, the ideology of mass science with heavy practical emphasis on close conformity with laws of production.

3. Science beyond the Gang of Four - marked by a renewed emphasis on reaching advanced levels of science and technology according to international standards, return to professionalization of scientists and scientific institutions coupled with ideological moderation, more attention to basic research.

These phases roughly reflect the dialectical 'contradictions and their resolutions' at the level of national policy and societal change. But they may not be considered strict negation of one another. If the theory of dialectics holds good for scientific and technological change, one would assume the present phase to have resolved the problems of the earlier phases while retaining the strengths gained through the process of conflict resolution itself. In more mundane terms it would imply that the Chinese science from now onwards would become quantitatively more vigorous and qualitatively more refined without losing its mass base, its socialist, people-centred approach. The recently announced eight-year science and technology plan further substantiates this line of argument. It is claimed that some 6000 scientists and technologists from all levels of activity and decision-making participated in the formulation of the plan which promises to accomplish the following long-term goals by the end of the century: help China reach advanced levels of scientific and technological capabilities according to international standards, equal encouragement and development of theoretical and applied research, full mechanization of agriculture and other sectors, completion of urban and rural electrification programmes, and general aid to building military capability. Even this cursory glance at the eight-year plan clearly indicates a blending of emphasis on building long-term scientific and technological capabilities as well as accomplishment of clearly specified technological goals at the national level. For the next eight years, the draft outline plan sets forth the following goals:

1. To approach or reach the advanced world levels of the 1970s in a number of important branches of science and technology.

2. To increase the number of professional scientific researchers to 800,000.

3. To build a number of up-to-date centres for scientific experiment.

4. To complete a nationwide system of scientific and technological research.

Advanced levels of theoretical and applied knowledge are to be reached in the following eight crucial areas, besides others: agriculture, energy generation through conventional and non-conventional sources, materials and natural resources discovery and development, electronic computers, laser technology, space research and exploration, high energy physics, and genetic engineering.[21]

The present situation in India regarding science and technology options in the national context is not quite clear. The period of the last Five Year Science and Technology Plan (1974-9) is just about running out. There has been no assessment so far of the relative success and failures in meeting the plan targets. There is no indication when the next plan document will be issued by the National Committee on Science and Technology. Issues regarding reorganization of science - particularly of the Council of Scientific and Industrial Research - have only been partially resolved. Policy position on such vital areas as energy is in the doldrums. The worst thing is that the new government has not come out with a definite statement on its position regarding science and technology research and development and their role in national reconstruction. Consequently, there is a deep sense of uncertainty and non-urgency, even apprehension, in scientific circles about the future of science in India and what the new leadership thinks or expects of science and scientists. On the positive side, one may cite the reformulation of the National Committee on

Science and Technology, the promise that the Committee will soon be coming out with a technology plan (the first in the country), and the issuance of the new industrial policy despite whatever shortcomings it may have. The most fundamental problem facing science planners in India today is twofold: to make the science and technology system operative toward producing small-scale technologies appropriate for the immediate needs of the common people, and secondly to bridge the gap between science and society, between science and production. Given the realities of Indian science and society, these are no mean tasks - but those that science must by necessity accomplish without wasting any more time. The Indian policy for science and technology, and the new model that would consequently emerge, must have the following five characteristics, besides others:

1. A clear demarcation of areas where technologies are to be obtained through international sources and no duplication would be allowed nationally.

2. A clear demarcation of areas where indigenous research and development capabilities, both theoretical and applied, would be built with utmost urgency to accomplish specific national goals within specific time periods.

3. A clear demarcation of institutions/institutional resources to be devoted exclusively to certain types of technological goals; e.g., rural technological research and development institutes with definite rural development goals as targets.

4. A clear enunciation of schemes through which all scientific and technical personnel could participate in actual developmental work from time to time, particularly in rural areas.

5. A clear demarcation of large-scale technological goals (for example, 'completion of low income rural and urban housing programme by the year 19..') to be accomplished through local resources and where necessary, both local and international resources.

Comparative Accomplishments and Mutual Relevance

As hinted earlier, the purpose of this work has not been to compare India with China for their performance in the field of science and technology, or development, on a one-to-one quantitative basis; although the perspective throughout has remained comparative. My purpose in this section is to make a few general remarks that may be of value toward a better understanding of relative accomplishments and failures of the two societies within their own peculiar circumstances and to comment on the relevance of their respective experiences for each other as well as for other developing countries.

In terms of comparative accomplishments in the fields of industry, agriculture, education, health, social welfare, cultural change - both India and China have had their successes and failures, their high points and low points. China may not have been able to completely eliminate the privileged class from society. A new privileged class of top party bosses and bureaucrats may have emerged. But in present-day China nobody is either very rich or very poor. According to authentic Chinese sources and outside observers, there is no malnutrition in China today, no slave labour, no exploitation of the poor by the rich, no large-scale human degradation of the kind known there for centuries. In India the masses are still poor. Some people are very very poor, some are very rich. The middle class has increased considerably in urban areas and to some extent even in rural areas, but is not sufficiently large to constitute a majority, as in other societies further advanced and organized on similar lines as India. Fruits of development are shared by a handful of people, mostly in urban areas. Children and adults suffer from malnutrition. Infant mortality is high. Sixty-five per cent of Indians cannot read or write. Yet India is an advanced nation in so many other ways. It has built an industrial, managerial, academic, scientific and technological capability unmatched anywhere in the Third World, including perhaps even China.

The world's reaction to India and China, as also the reaction of the two nations to each other, has been negatively or positively biased and very often quite distorted. For instance, when America ignored China for two-and-a-half decades after the revolution, the rest of the world (except India and Russia) also seemed not to care much about China. All reaction was by and large negative. The human cost of the Chinese experiment was considered too high. The Chinese had violated basic norms of human decency by forcing people to give up certain things and do others, even if they had accomplished certain practical results on a massive scale - or so it was argued. Today it is China all over, thanks largely to Richard Nixon's 'discovery' of the Great Wall. Everyone (excluding India and Russia) seems to be so much enamoured by the 'Chinese experience'- all those great accomplishments of mobilizing resources, motivating people on a scale unsurpassed in history, being rational and pragmatic. Nobody seems to care much to talk about the continuing denials of basic freedoms, the hard-core psychological attack on the human mind, the unpersonification of persons, etc.

India's failures: her famines, beggars, maharajas, snakes and cows are so much better known and talked about than her accomplishments, thanks largely to the Indians' own endless propensity to criticize publicly and downgrade themselves. People everywhere criticized India for allowing too much freedom to her own people, for not implementing plans, for being a soft state. When freedoms were suddenly taken away by Indira and Sanjay Gandhi, everyone said those were India's greatest assets. Now, when democracy and the due process have been restored, people are saying Indira Gandhi and her son have not been dealt with properly for their crimes. The fault, I suppose, is primarily ours: we don't do anything right - and in right earnest.

Unlike India, everything in China - economy, science and technology, culture and social relations, even droughts and earthquakes - is political. Earlier people everywhere used to say how silly it was of the Chinese. Now people everywhere seem to be saying that is the greatest thing they do in China: they politicize everything and thus generate a necessary commitment to programmes and motivation to work. What can a society like India learn from such experiences? Indian intellectuals have often asked if India could do better with this approach; to which the answer has invariably been that India has not had a Chinese type of revolution and socio-economic reorganization and that without such revolution and reorganization, politicalization of issues and mass mobilization would not be possible. What this answer implies in effect is that it has got to be either wholesale or nothing. To my mind, the issue cannot be closed that easily and abruptly. India and China both can learn a great deal from each other by sharing scientific and technical knowledge they have acquired to handle the problems of industrial and agricultural production and their distribution to the masses. The Chinese formula for distribution purposes is so much more successful than anything India has been able to accomplish. It is true that production and distribution functions cannot be independent of the social and political conditions in a society - as Marx said. But the path toward building a socialist society need not be entirely Marxian either; or does it have to be? India has had Gandhi and Jayaprakash Narayan who have found little, if any, expression in India's policies for large-scale development and social change. While the Chinese developmental model has many unique features, in its entirety it may not be so exclusive as it is often made out to be by Western China watchers who know something about China but know relatively little or nothing about India. For instance, the notion of a non-consumptive, non-market rural economy based on agriculture and rural industry immediately brings Gandhian economics into sharp focus. Of course, as J.D. Sethi rightly points out in his comparison of Mao and Gandhi, the Gandhian model was never put into practice. Its strengths and weaknesses were never tested in practical terms.[22] Unlike Mao, Gandhi never ruled over India and never had a chance even to influence the Indian nationalist government to try his ideal social and economic organization at least on an experimental basis. On the contrary, the Indian planners considered the Gandhian model as reactionary and revisionist - a threat to Indian modernization - and as such, rejected it totally. It is only in posterity that we begin to realize that no model, either Gandhian or Maoist, is complete and perfect and need therefore be accepted or rejected in totality; that perhaps the Gandhian formulas were quite suitable, at least for a limited period of time, to reduce, if not completely eliminate, problems of rural production, poverty, unemployment, and concentration of economic and political power in India; and that Gandhian economics, like small rural industries in Mao's China, may not necessarily have clashed with scientific and technological progress, mechanization of large systems of production and modernization of society at large. By ignoring the Gandhian model altogether India may have missed a unique opportunity of trying something uniquely Indian which could have relevance for India as well as the rest of the world.[23]

Commenting on the relevance of the Chinese model of science and technology for other developing countries, Susan Rifkin declares categorically: 'The model is inappropriate in this context for several reasons.'[24] The most basic reason for this inappropriateness, she says, is the uniquely Chinese nature of their political, economic and social structure unmatched anywhere in the developing countries. This is the 'either you can take it all or nothing' approach pointed out earlier. Indian planners and elites would also say 'since we cannot take it all, we can take nothing'. If that were to be so, very few societies, systems, or people in the world could learn anything from others. Another reason for inappropriateness is lack of a large labour force and the appropriate political and economic structure for its full-scale participation in national developmental activity. It is indeed true that the Chinese mobilization strategy has played a signal role in that country's rural transformation and that other societies may never be able to accomplish anything quite like that.[25] However, the first condition of a large labour force is met in the Indian case. The problem of mobilization of this labour force remains. India does not have the structure of work, leadership, ideology, etc. of the Chinese type for this purpose. But I do not see any reason why communities on a piecemeal basis cannot be mobilized by the leadership with the help of other elite groups, such as scientists, social workers, community leaders, administrative personnel, professionals of all varieties. Such mobilization efforts on an *ad hoc* basis are not uncommon in India. The problem lies with the Indian approach and philosophy where development is done for the people by specialized agencies without involving the recipient communities. Instead of creating additional bureaucratic systems to undertake new tasks, the Indian approach must move toward community-based, institutionalized mechanisms for support building and mobilizing people to do the work themselves. Someone in the community - political leaders, professionals, military and police, and others - have to take the lead. India has developed no tradition and consequently no expertise of this type of community action at any level. But it can be done. Susan Rifkin's other reasons for making the Chinese model inappropriate for developing societies are large internal markets in China, but absent elsewhere, and the 'ability of the Chinese to adapt science and technology . . . coupled with community responsibility (providing) a base for science and technology that does not exist in most developing countries.' Actually, these are some of the very reasons why the Chinese and Indian models can be so relevant to each other. India has a vast potential in rural areas, 'markets' which Indian industries have not even begun to tap in their mad rush toward export markets. These markets could be an inexhaustible source of demand for inexpensive technologies and consumer products, both of which can be procured and produced through totally indigenous resources. India, like China, has had a long tradition of scientific work and technological innovation.

The myth of the conservative Indian who refuses all innovation has long been broken in the countryside, especially in the process of adoption of modern agricultural technology. And finally, India, like China, is also a communal type of society ... a *gemeinschaft*. The community spirit manifests itself in countless cultural forms and practices. The problem lies somewhere else: the Indian leadership and the Indian developmental administration have failed to use that spirit for purposes of social and economic transformation.

India and China: Global Implications

The most important global implications of India and China rest primarily on two factors: (1) nearly one third of all humanity belongs to these two countries; (2) both of them have acquired scientific-technological and military capabilities of significant magnitude.

India has always been open to world markets and technology, sometimes to her own detriment. India's role in the international techno-economic order is going to increase considerably in coming years. Indian expertise is already being exported in sizeable measures to other countries of the Third World, especially in the Middle East. Export of manufactured items has also gained momentum in recent years to the point where India for the first time in her economic development has a favourable balance-of-payment position. This has been possible by a long-standing strategy of import substitution in the industrial sector, considerable increase in agricultural production during the last decade which has reduced the need for foodgrain imports, and various other measures to conserve foreign exchange. Agricultural research has by now established a tradition of making substantial contributions toward improved farm practices leading to higher yields. Indian industry is for the first time responding to the need for indigenous, in-house research and development as an additional measure of building self-reliance. The governmental research and development structure is being currently reorganized to make it more sensitive to the needs of society. Naturally, all of these measures will have some successes and some failures. But in any case, given the necessary political stability and leadership at the central and state government levels, India should make very rapid strides toward emerging as a viable self-reliant society within the next two decades to acquire a prominent position among the industrialized nations of the world. However, problems of overpopulation, poverty, malnutrition, and various other types of deprivations for the masses will remain - unless the age-old structure of vested interests, inequality, and exploitation is removed by redistributing wealth and related benefits among the weaker sections of society by means of effective social legislation and welfare programmes. If the Indian masses, or for that matter masses anywhere, remain poor and hungry, it will have repercussions for the entire world. If hunger has to be removed quickly and effectively from the world, rich nations will have to give up some of the luxuries of super-abundance, do something more concrete than show interest, pity and curiosity in the condition of the poor and the deprived, and give them something more than expert advice - not as charity but as moral responsibility. This would involve transfer of world capital and technology on a global basis, from high-capital, high-technology areas to low-capital, low-technology areas under an international safeguard system protecting the interests of both the receiving and supplying nations.

China has already gained a commanding position in the Third World. Due to her growing need for advanced technology and export markets, it would be foolish not to expect China to open up more and more to the rest of the world. The new Chinese leadership seems to be fully aware of the fact that isolationism of the past variety will not lead China toward an equal status in the community of nations, and that if China is to become a world power industrially and technologically - if not necessarily militarily - she will have to interact more with others on a give-and-take basis. My forecast is that China will be fully integrated in the international economic order by the turn of the century. This carries the possibility of opening up vast Chinese markets, at least gradually, to other manufacturing nations of the world. Finally, and from the point of view of this analysis most importantly, this also involves the possibility of Indian-Chinese collaboration on a number of techno-economic development fronts, particularly in rural and management development schemes, programmes for alternative sources of energy conservation and generation (including nuclear energy for limited peaceful purposes) and development of alternative technologies for a vast variety of tasks.

Lastly, I would like to raise a question which has so often been raised: Do the Indian and/or the Chinese model(s) of development offer meaningful alternative(s) to a fast-changing world? Naturally, I do not propose to provide an answer to this extremely complex question through an elementary analysis of this type. One obvious thing, though, is that in the Third World framework the case of Japan must never be forgotten. China and Japan may be considered two extreme opposites in their political and economic systems and policies for scientific and technological and related development. No two countries in such comparisons can be considered as pure types - but while Japan has emphasized the philosophy of 'borrowing and adaptation' from outside, mostly from the West, China has always professed the philosophy of self-reliance through indigenization. Given her own needs and historical realities, Japan's is the most successful success story of the Third World. India is a mixed case - something in between China and Japan. Her gains and losses can also be viewed as mixed. Given her experience of the past thirty years, it would be naive to think or suggest that India can change to either of the 'pure' types. But we must expect India to be certain and know where and when it would like to go the 'Japanese way'[26] and where and when it would like to do 'like the Chinese' and to come out with a unique blend or formula of her own that could lead to the third success story of the Third World and provide another alternative to others climbing up the developmental ladder.

The question of alternatives is far deeper than techno-economic formulas, however. It is a question of goals and purposes of societies. It is a question of values, of styles, and what may constitute the quality of life. Is it not that Japan, India, China and other developing societies may all be seeking the material utopia of the West reached through technological and politico-military competence? If that is the way it is, then India and China (and Japan) do not offer an alternative to anyone in the true sense; just alternative forms of socio-economic and technological organization and management to reach identical goals. In that case there would be many more Vietnams, Biafras and Bangladeshes and supertankers full of oil would keep spilling pollution on the shores of civilization.

NOTES

1. The uniqueness of the 'Chinese model' should not be accepted without careful scrutiny. Similarities and differences between the Chinese approach and those adopted in other countries for solving similar problems must be examined in tactical and strategic terms not just at the level of larger political and economic systems.
2. Sarwar Lateef, 'Economic Growth in China and India - 1950-1980', Vol. 1, *Economic Intelligence Unit Report* No. 30, September 1976.
Subramanian Swamy, 'The Economic Distance Between China and India' *The China Quarterly,* No. 7, June 1977, pp. 371-82.
3. Rushikesh Maru, 'Research and Development in India and China', Unpublished draft report. Research Policy Program, Lund, and Centre for the Study of Developing Societies, New Delhi, 10 April 1969.
4. See, for example, the following studies of Jon Sigurdson: 'Development of Rural Areas in India and China', *AMBIO,* (Journal of the Royal Swedish Academy) Vol. 5, No. 3, 1976, pp. 89-107; 'Water Policies in India and China', *AMBIO,* Vol. 6, No. 1, 1977, pp. 70-76; 'Rural Industrialization: A Comparison of Development Planning in China and India', *World Development,* May 1978.
5. Patricia A. Vaughan, *Science and Science Policy in India and China: A Comparative Approach,* B.A. Honours Thesis, Department of Political Science, University of Melbourne, 1976.
6. See, for example: Ward Morehouse, 'Professional Estates as Political Actors: The Case of the Indian Scientific Community', *Philosophy and Social Action,* Vol. II, No. 4, October-December 1977, pp. 61-95; Aqueil Ahmad, 'Management of Human Resources in Science: The Indian Scene' *Indian Journal of Public Administration,* Vol. XXIII, No. 1, January-March 1977.
7. The examples of IBM India and the Coca Cola Bottling Company may be most instructive in this respect.
8. Most of the data on the behaviour of multinational corporations and other foreign firms in India presented here has been taken from various studies reported in *Current,* Special Issue on Multi-nationals, 26 January 1978.
9. Edgar P. Snow, *Red China Today: The Other Side of the River,* Harmondsworth, 1970.
10. K.K. Subrahmanian and P. Mohanan Pillai, 'Implications of Technology-Transfer in Export-Led Growth Strategy' *Economic and Political Weekly,* Vol. XI, No. 44, 30 October 1976, pp. 729-34.

Peter Neersø, 'Some Aspects of India's Policies on the Import of Technology', *Development and Change,* Vol. VI, No. 1, January 1975.
Both of these studies have arrived at similar conclusions, but are silent on the causative factors leading to the present policies on technology transfer or the alternative modes of creating technological foundations for a growing economy.

11. K. Jayaraman, 'International Investment and Multinations Inevitable for Developing Nations', *Current,* Special Issue on Multinationals, 26 January 1978. Jayaraman's work fails to take into consideration negative aspects of multinationals and other foreign firms on the Indian economy, particularly the fact that their contribution to rural development may have been inconsequential, if not entirely negative.
12. For this genre of literature, see Barry Richman, 'Chinese and Indian Development: An Interdisciplinary Environmental Analysis', *American Economic Review,* Vol. 65, No. 2, May 1975, pp. 345-55; Thomas Weisskopf, 'China and India: Contrasting Experiences in Economic Development', *American Economic Review,* Vol. 65, No. 2, May 1975, pp. 356-64.
13. A brief analysis of this problem is presented in Aqueil Ahmad (Ed.) Introduction to *Management of Agriculture Research in India,* forthcoming.
14. For a critical commentary on India's New Industrial Policy see: C.T. Kurien, 'Small Sector in New Industrial Policy', *Economic and Political Weekly,* Vol. XIII, No. 9, 4 March 1978, pp. 455-61; K.K. Subrahmanian, 'Approach to Foreign Collaboration - A Critique of New Industrial Policy', *Economic and Political Weekly,* Vol. VIII, No. 14, 8 April 1978, pp. 613-17.
15. Similar doubts have been expressed by Genevieve Dean in *Science and Technology in China,* Lecture Series -4/77, The Centre for the Study of Science, Technology and Development, CSIR, New Delhi, 1977.
16. For a further commentary on 'mass science' notion in China, see R.P. Suttmeier, 'Party Views of Science: The Record from the First Decade', *The China Quarterly,* No. 44, October-December 1970, pp. 146-68.
17. For a detailed discussion of the Chinese strategy of making science relevant for the common people, see *China: Science Walks on Two Legs,* a report from Science for the People, Inc., Avon, 1974.

18. Despite current shifts in the Chinese science policy, traditional links between science and production are not only likely to continue but to be further strengthened:
'The revolutionary teachers have lucidly explained the dialectical relationship between science and production, which is to say, science originates from and at the same time gives guidance to production.'
(Chung Ko, Scientific Research Should Precede Production - A Criticism of the 'Gang of Four's' Distortions, *Peking Review*, No. 27, 1 July 1977, p. 13.)
19. For an explanation of the 'limbo effect' in Indian Science, see Aqueil Ahmad, op cit. No. 14.
20. The denigration of the 'Gang of Four' for over-politicization of science is a common feature of the current official Chinese position on the subject. See, for instance, 'Gang of Four' - 'The Nation's Scourge', *Peking Review*, No. 27, 1 July 1977, pp. 14-17.
21. 'Outline National Plan for the Development of Science and Technology, Relevant Policies and Measures', *Peking Review*, No. 14, 7 April 1978, pp. 6-14, 17.
22. J.D. Sethi, Mao and Gandhi, 'Convergence and Divergence - A Preliminary Note', *China Report*, Vol. XIII, No. 1, January-February 1977, pp. 16-33.
23. An elaboration of this line of argument is provided in Aqueil Ahmad, 'Dilemma of A Developing Society: India', *Futures*, Vol. 9, No. 3, June 1977.
24. Susan B. Rifkin, 'The Chinese Model for Science and Technology: Its Relevance for Other Developing Countries', *Development and Change*, Vol. 6, No. 1, June 1975, pp. 23-40.
25. For an explanation of the Chinese strategy of mass mobilization for economic development, see Robert C. Hsu, 'Mass Mobilization as a Strategy of Economic Development: The Case of Chinese Agriculture in Yenan Period (1936-45) and Post-1949 Period', *Weltwirtschaftliches Archiv* (Review of World Economics), Band 111, Heft 1, 1975, pp. 164-75.
26. The question of relevance of the 'Japanese Model' for India has been briefly dealt with in A.K.N. Reddy, 'Basis of the Japanese "Miracle": Lessons for India', *Economic and Political Weekly*, Vol. VII, No. 26, 24 June 1972, pp. 1241-45.

LOCAL TECHNOLOGY SYSTEMS IN INDIA AND CHINA

Ward Morehouse

Marginalization of Science in the Third World

It has by now become commonplace to observe that formal science and technology systems in the Third World countries have a marginal relationship to the most pressing economic and social problems of those countries. These systems, composed of national science councils and other agencies, basic and applied research laboratories, and the science and technology components of the tertiary level of the educational system, have been built up over the past 30 years, especially in the industrially more advanced developing countries, in the expectation that they would trigger a social and technological transformation of these countries.

That this has not occurred does not mean that these systems have been total failures. Even though their capacity to generate 'break-through' technological innovations along the lines associated with research and development systems in the larger industrialized countries has been virtually nil, they have in fact been an important source of skill formation and diffusion of technological competence in the modern sector of the economy and have served substantial national security objectives in those Third World countries which have created the largest such systems in the Third World, namely India and China.

The reason why these formal science and technology systems have failed to trigger the social and technological transformations essential to the widespread alleviation of poverty in these societies seems almost simple-mindedly self-evident in retrospect. They are alien cultural imports, modelled after systems nurtured in entirely different social contexts. Being derivative in design and foreign in conception, they have remained in many ways peripheral structures dominated by the scientific metropolises in the centres of political, economic and military power in the international system. The assignment of such transformational tasks to these systems is as unrealistic as expecting the Bolshoi Ballet Company and New York Philharmonic Orchestras, separately or together, to put a man on the moon in three years or thirty. The goal is unrealistic and inappropriate.

Figure A
Linkages of R and D Systems to Contemporary Societies

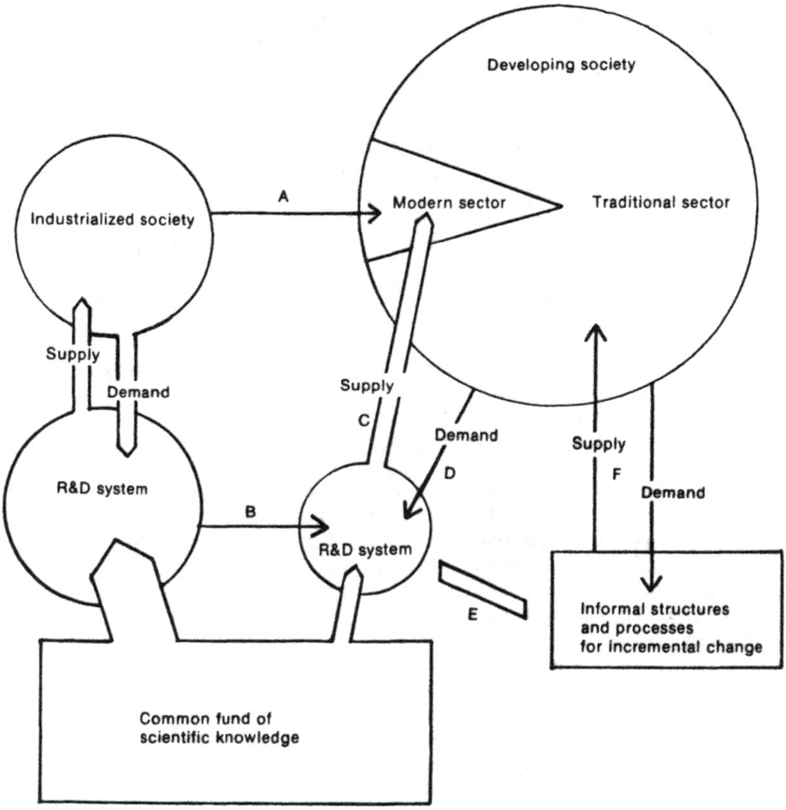

While supply and demand between the R&D system and the economy are in rough equilibrium in industrialized societies (left), this is not the case in the developing world (right). Industrialized societies determine consumption patterns (A) in the modern sectors of developing societies, and R&D systems in industrialized societies define the research tasks (B) in the R&D systems of developing societies. The R&D supply to developing societies (C) is mostly limited and goes only to the modern sector. The R&D demand from developing societies (D) is weak. There is a limited supply and demand between developing societies and their informal structures and processes for incremental change (E). There is virtually no interaction between the formal R&D system and those informal structures (F). Industrialized societies provide 90 to 95 percent of the sources of technological change in developing societies.

Source: Ward Morehouse and Jon Sigurdson, 'Science, Technology, and Poverty: Issues Underlying the 1979 UN Conference on Science and Technology for Development,' *Bulletin of the Atomic Scientists,* December 1977

Figure B
Linkages of R and D Systems to Comtemporary Societies:
A Hypothetical Model for the Future

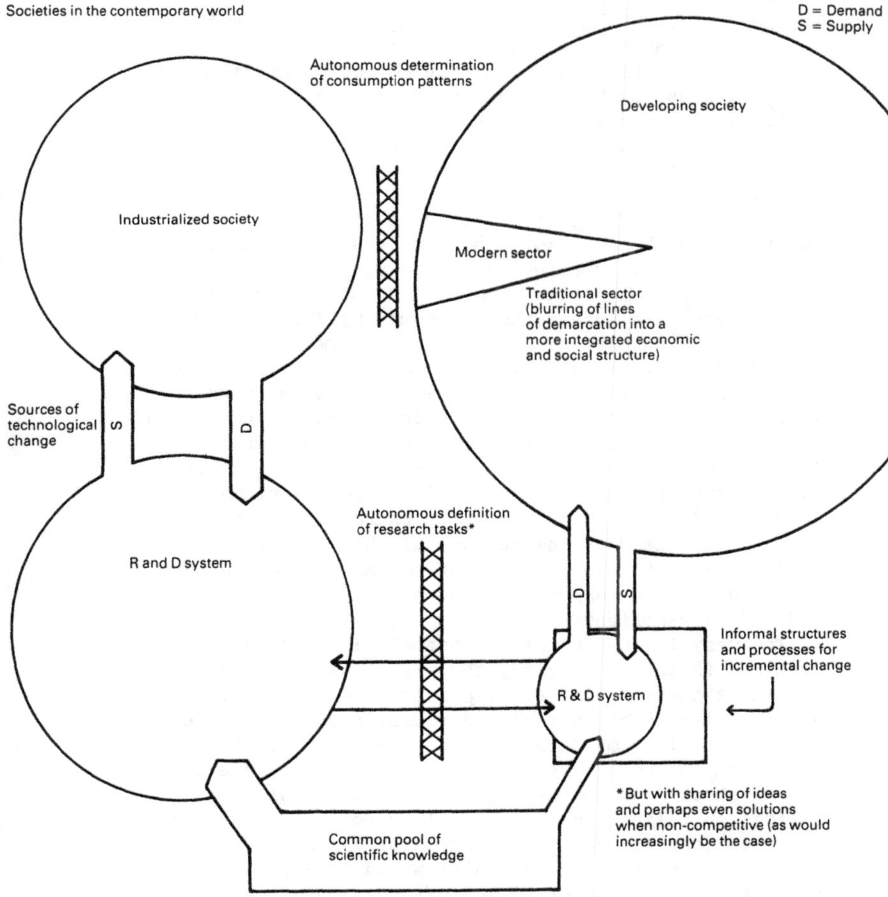

Note: This model attempts to represent only the relationship of sources of technological change to contemporary societies and does not reflect the many other interconnections and interactions between industrialized and developing societies which are manifested in common global problems shared by all mankind.

Source: Ward Morehouse, 'The Endless Quest: Science and Technology for Human Betterment in India,' *Research Policy Program Discussion Paper Series,* Lund: University of Lund, December 1976

The nature of this dysfunctional relationship between a formal research and development system in a developing society and the needs of most of the people can best be represented graphically in the accompanying diagram of the linkages of research and development systems to contemporary societies (Figure A). It is possible to construct a hypothetical model of how these systems for technological change might be more functionally linked to developing societies so that the systems would better serve the needs of most of the people (see Figure B). Assuming for the moment that science and technology have a significant role to play in the social transformation of poor societies, the critical question becomes how to move from the situation portrayed in Figure A to that pictured in Figure B. It should be emphasized that no Third World society has yet achieved that latter situation which remains a normative model.

Some would argue that among the larger Third World countries which have made the most substantial effort to develop a formal science and technology system to serve the modern sector of the economy and national security needs, China has gone farthest in moving from A to B. Certainly China has, as best we can determine, made a considerable effort to develop local technology systems to serve rural productive sectors of the economy which involve most of the population. In the Indian case also, considerable effort has been made to deliver technological inputs to these productive sectors but without constructing the kinds of local technology systems based on mass participation which characterize the Chinese experience. The purpose of this introductory paper will be to explore some of the methodological and conceptual issues growing out of a study of the Chinese and Indian experiences in developing such systems.

The very limited number of actual cases being examined in both countries, the preliminary nature of the investigation, the still poorly understood relationship between technological and social change, and the obvious problems of data collection and verification in China and India ('jigsaw puzzle' versus 'needle-in-the-haystack' scholarship) provide little basis for making generalizations and even less for setting forth prescriptions for designing such systems in other countries elsewhere in the Third World.[1] Hence the character of this paper as an introductory comment on some of the methodological and conceptual issues arising from such an on-going study.

Local Technology Systems

Poverty and Technological Change

It is sometimes asserted that the failure of most Third World societies, including India, to satisfy the minimum physical requirements of food, clothing, and shelter of substantial numbers of their people, let alone the equally basic non-physical needs of productive employment, education, good health, and psychological self-realization, is causally linked to the marginalization of modern science and technology in those societies. Indeed, some critics of science and technology establishments in poor societies go farther and argue that the persistence of poverty is in some measure attributable to the marginalization of modern science and technology because of the diversion of creative human talents and scarce financial resources to serving competing needs of the small, largely urbanized modern sectors of these societies while neglecting the poverty-stricken, primarily rural majorities.

With technology as a dependent variable in the process of social change, the causal link appears weak, at least in the negative sense of preventing the alleviation of poverty.[2] On the other hand, the Chinese experience, to the extent we understood it, suggests that the process of 'technological mobilization' can be a significant factor in bringing about the social transformation involved in meeting the most essential material needs of most, if not all, of the people.[3]

To understand the role of technology in social change, we must examine the process of technological change itself and how it interacts with changing economic, social and political relationships and institutions. Analysis of the process of technological change should, ultimately if not immediately, seek to shed light on a somewhat re-formulated version of one of the classic definitions of politics:

> Who gets what (including how much) when, where, why, and how with the introduction of a technological creation or acquisition (or adaptation of an existing technology) into an established social unit (family, community, enterprise, province, region, nation or group of nations)?

Answers to these questions are crucial in devising strategies which involve using technology as one instrument for the alleviation of poverty. Failure to pose such questions is one reason why technological interventions which have the stated purpose of helping the poor do not work. If they are unaccompanied by a reconstruction of economic and political power within the community (or by a sustained external intervention of power to protect the weak), and threaten to erode the position of dominant elements in that community, the fruits of the technology, if not the technology itself, are likely to be appropriated by those stronger elements from the most deprived.

It is by now well established that external technological interventions can have substantial impact on established social units. The elaborate analyses undertaken by the economists of the Green Revolution make a conclusive case.[4] A very different kind of case is made by Claude Alvares in his provocative re-examination of the history of technology and social change in India, China, and the West from A.D. 1500 to the present. He shows how the extension of the technologies and productive systems of the industrialized countries into the Third World is displacing the poor from their land and existing occupations and destroying traditional technologies and productive systems.[5] Both, however, make in different ways the same point: that without a reshaping of power within the social community, technological change is unlikely to alleviate poverty and is much more likely to exacerbate inequality within the Third World.

A Local Technology System Defined

Another reason why such interventions do not work, at least on a sustained basis, is the failure to devise an appropriate supporting technology system within the community into which the technological intervention is being made. This system may or may not be coterminous with the social unit most directly concerned. In the case of international transfer of technology, it almost certainly will not be, as the boundaries of the system must be extended to include the source of the technology, and the source must be analysed in terms of the imperatives for transfer of technology which operate on it.

A technological system may or may not include formally organized institutions for research and development. In Third World societies, it will often not include formal research and development institutions since much technological change appears to occur within the context of local productive systems such as small-scale industrial enterprises and the farms of individual or small groups of farmers making incremental improvements in existing technologies through empirical observation and trial and error. These technologies are then diffused to other users through word of mouth and demonstration or migration of workers from one productive enterprise to another, who carry the innovation with them in their heads. Sometimes, there are formally organized extension mechanisms under governmental auspices, sometimes not. In industrialized countries, formal research and development institutions are much more likely to be involved in technological innovation.

A technology system is thus a set of interrelated institutions and processes which generate the essential technical skills and knowledge for a sector of productive activity or social service within a society.* A local technology system with informal and formal organizational components more closely linked to productive activity and greater user participation is more likely to serve the needs of those sectors largely or entirely neglected by formal science and technology systems. In most Third World countries which have developed formal science and technology systems on any scale, those sectors largely or entirely neglected by this formal system include subsistence agriculture, small-scale industry, and the life-support goods and services which those working in and supported by these productive sectors need or should have even if they do not always get them - for instance, food, health care, clothing, shelter, energy and transportation.[6]

A local technology system might serve any community and/or productive or service sector, including villages and groups of villages or a district town and surrounding rural community. By way of illustration, the structural components and activities performed by these structures in a local technology system serving a group of villages and linked to regional and national levels is outlined in Figure C. Some elements in this kind of system already exist in many Third World countries, typically the upper levels in a hierarchical or scientific sense at the provincial or national levels, but only rarely does the structure extend down to the local level of

---

* A good working definition of technology for my purposes is that given by the *Encyclopaedia of the Social Science:* The skills, tools, knowledge, and procedures for making, using and doing useful things ('things' being understood to include services).

Figure C
Outline of a Local Technology System Linked to the Provincial and National Levels

Level of productive activity or administration	Research and dissemination structure	Activities performed by the structure
Single village	Innovation and dissemination group (composed of regular farmers or workers with limited special training)	Testing of innovations generated elsewhere in the system; training other farmers and workers in their use; feeding local improvements and refinements into the system
Group of 10–15 villages	Research station (small regular staff plus local farmers and workers from innovation groups on rotating assignments)	Experimental and seed breeding plots to test/generate new varieties; dissemination of information to local innovation groups and farmers/workers
Sub-district or development block (200–500 villages)	Research centre (larger regular staff, some with technician-level training; local farmers and workers on rotating assignments)	Experimental and seed breeding plots; design of agricultural implements for local fabrication; experimentation with new cultivation practices; training of local innovation group personnel
County or district	Agricultural research institute; agricultural machinery development centre (staffed by those with technician-level training; farmers/workers from other levels of the system on rotating assignment)	Testing of varieties developed at provincial and national level; testing of prototypes of machinery; experimentation with methods of irrigation, moisture control, pest control, use of fertilizers, short training courses for personnel from research stations and centres within county or district, etc.
Provincial and national	Agricultural universities, specialized research institutes on different crops and agro-industries producing inputs for agriculture such as fertilizers and pesticides (university graduate scientists and engineers, technicians, local farmers/workers as part of *ad hoc* development groups, etc.)	Advanced training and research, including development of new varieties of seeds, fertilizers, pesticides, implements, production processes for these inputs, etc.

Source: Carlos Contreras, Charles Edquist, Osita Eze, Ward Morehouse, B. V. Rangarao, and Mervyn Wijeratne, *Technological Transformation of Developing Countries* (Paper based on informal consultations organized by UNCTAD, Geneva, January 1978), *Research Policy Program Discussion Paper Series No. 115,* Lund: University of Lund, February 1978.

production. And even more rarely is there active participation of farmers and workers in the innovation and dissemination process and multi-directional traffic horizontally (to other productive units at the same level of activity) and vertically in both directions, involving people, ideas, and experiences.

Among the principal characteristics of a local technology system designed to alleviate poverty and help meet the most important material needs of most, if not all, the people in Third World societies are these:

1. Close links to, and in some cases, being an integral part of the productive system.

2. Active participation of users in problem identification and the testing of solutions.

3. Emphasis on rapid diffusion of tested innovations.

4. Existence of feedback mechanisms so that user refinements and improvements can be quickly incorporated into redesigned innovations and in turn widely disseminated.

5. Attention to small, incremental improvements in existing technologies as well as to 'breakthrough' innovations.

6. Where new solutions to fundamental problems are being sought, the existence of a strong scientific base for the necessary research.

The foregoing are characteristics of a model of such a system. An illustration which comes close to the model will be found in the case of a Chinese commune being examined in this study. How widespread this type of local technology system is throughout China is difficult to say, although there is some indication that it is fairly well established among many, perhaps most, rural communes which carry on most of the agricultural production and a substantial volume of small-scale industrial activity and where most of the population of the country lives. Virtually all of these elements will be found in India in isolated instances, some of which are also being examined in the study, but they do not yet appear to have been put together in a systematic way, let alone made the basis for any kind of comprehensive approach to the alleviation of rural poverty and satisfaction of the minimum needs of most of the people.

Ward Morehouse

Studying Local Technology Systems in India and China: Problems of Data Collection and Analysis

This introductory paper on methodological and conceptual issues grows out of a larger, on-going study to examine local technology systems in India and China. The central thrust of that larger study is to increase understanding of structures and processes for technological change in the alleviation of poverty and satisfaction of minimum material needs of most, if not all, of the population of these countries. Since the proportion of the population in both countries in urban areas is relatively small (roughly 20 per cent), this means an inevitable focus on rural society.

Case studies of efforts to build local technology systems in rural areas have been, or are being, undertaken in both countries. In China, principal attention has been given to a commune in Guangdong Province, Huadong by name, which I visited in 1973 and about which I have published a paper seeking to look at that commune as a technological system ('The Commune as a "Technological System": Notes on Hua-tung Commune', *China Quarterly,* September 1976, pp. 582-596). An attempt is being made to place commune-level activity to generate technological change in a larger regional and provincial context, with particular reference to the work done by Jon Sigurdson in Hebei Province and reported in his *Rural Industrialization in China* , (Cambridge, Mass.: Council of East Asian Studies, Harvard University, 1977).

In India, three case studies are initially being undertaken, although other cases of rural technology development projects may be added. One is a programme in the state of Gujarat to increase the productivity of small farmers through improved agricultural technology using non-formal education techniques. This programme has been undertaken by the Behavioural Science Centre at St. Xavier' College in Ahmedabad; some of the early experience in the Centre has been described by the originator of the programme, Professor J.M. Heredero, in his *Rural Development and Social Change: An Experiment in Non-Formal Education,* (New Delhi: Manohar, 1977).

The second case is that of an organization at the state level striving to build a 'mass science' movement in the rural areas and seeking to use the movement as a means of applying scientific knowledge and new technological methods to improving the lives of rural communities. The name of this organization is the Kerala Sastra Sahithya Parishad which functions in one of the most densely populated but socially most developed states in India - the state of Kerala in South India.

The third case is an effort to tap the roots of India's technological traditions in solving rural social problems. Located in the state of Goa in Western India, the Rural Study and Transformation Centre (RUSTIC) plans not only to analyse Goan rural problems and identify or develop technological and socio-economic solutions but to implement them, thus becoming an instrument of social change in the Goan countryside.

These initiatives are but three of literally dozens, if not hundreds, of rural development schemes undertaken in India in recent years by universities, scientific research institutes, business enterprises, and voluntary social action organizations to alleviate rural poverty through improved methods of productive activity. Few of them have been systematically studied and externally assessed. In these circumstances, the problem of selecting cases for study becomes crucial. Since these cases are being studied for the most part in real time, repeated access over extended periods of time is equally crucial.

Repetitive access over time is essential to understanding the dynamics of any social system, but unfortunately in China, the opportunities for this repetitive access are much more limited. It would be extraordinarily useful for me to be able to return to Huadong Commune six years after my initial visit to study the changes which have occurred; such a second visit appears to be remote. There are many other constraints on data collection in China growing out of the still very limited and controlled access to the country by foreign scholars and the great paucity of published macro-economic and social data, resulting in the phenomenon of 'jig-saw puzzle' scholarship to which reference has already been made. It is difficult in such circumstances to place local cases in a regional or national context, to know how representative or exceptional they may be, and to compare local performance in different parts of the country.

While India is by contrast an open society with such an abundance of published information that one encounters the opposite problem of screening out of a mass of material the reliable and relevant data ('needle in the haystack' scholarship), in both countries very little data exist on the social impact of technological change, especially over time, whether at the micro or macro level. Such data are not, furthermore, easily collected even where a researcher does have repetitive access over time. And there is finally the obvious limitation of attempting to generalize from a few local or even regional cases about two continental-sized countries with enormous variations within them.

The Problem of Conceptualization in Studying Local Technology Systems

The central question in such an inquiry is whether there is validity to the concept of a 'local technology system' as a framework for data collection and analysis. Can a 'technology system' be distinguished from a productive system or a governmental/administrative system? Are such systems coterminous or do they have distinctively different scopes? Is it possible for a technology system to exist independently of a productive or governmental/administrative system? Would it be more useful simply to examine the role of technology as a factor in change within some established social system?

While I do not profess to have final answers to these questions my working hypothesis at present is that there is utility, for instrumental if not intellectual reasons, in attempting to employ a local technology system as the basic framework for data collection and analysis. It is necessary first to make a distinction between a technology 'system', implying some measure of institutionalization of structures and processes for generating and diffusing technological change, and largely informal social processes and interactions leading to technological change. The presumption is that institutionalized processes and structures, with explicit form and content, are more susceptible to being introduced into a rural community or region through deliberate and conscious intervention, whereas informal processes and interactions are much less susceptible to stimulation through such interventions or are less likely to be self-sustaining without some kind of supporting structure.

If the goal of generating technological change in rural areas in countries like India and China is to alleviate poverty and meet the minimum needs of most of the population of these areas, then deliberate and conscious interventions seem to be essential. The historical evidence of the past 30 years suggests that in China, where such deliberate and conscious interventions have been (and still are being) made on an extensive scale, the productive capacity of rural areas has been significantly increased and poverty substantially alleviated although hardly eliminated. The basis may have been laid, furthermore, for a more sustained and rapid future growth in productivity and productive capacity in the Chinese countryside in the years and decades immediately ahead. Critical in the Chinese experience has been the linking of the development of local technology systems to basic economic and social changes such as the collectivization of agriculture and the elimination of landlords and rich peasants as a social class in the countryside.

In India by contrast, where little effort has been made to build local technology systems closely linked to local productive activity and with the active participation of farmers and workers at the local level, poverty remains a widespread phenomenon, with studies estimating that 40 per cent of the Indian population live below the Indian 'poverty line'. Where there have been interventions to generate technological change, they have generally been made within the framework of the existing social order in rural communities, thereby reinforcing or even exacerbating the inequalities within that order. Even in areas where substantial economic growth has occurred and informal processes for technological change appear to have been a key factor such as the Punjab, rural poverty remains a significant social problem. Without building local technology systems explicitly directed toward meeting the needs of the most economically and socially disadvantaged segments of rural society, it appears difficult to alleviate poverty and meet minimum material needs on a self-sustaining basis - i.e. through increasing the productivity and productive capacity of the poor.

It is nonetheless clear that we need to know much more about not only the structural character of such technology systems but also their dynamics - i.e. how they function over time and with what social consequences. And we need careful, systematic studies which seek to compare the socio-economic efficacy of institutionalized systems with informal processes and interactions resulting in technological change.

Theoretical and Instrumental Issues in Studying Local Technology Systems

The relationship of social to technological change is extremely complex, and while extensive studies have been made of their interaction in a variety of social, cultural and historical contexts, few tested generalizations with the capacity to predict with reliability the course of change in other contexts exist. A basic theoretical issue is the nature of the relationship of technology to economic, social, political, and cultural variables in the process of change. While I hold to the position that it is generally a dependent variable (save in special circumstances such as advanced military technology) in the sense that it is unlikely to be the sole engine for generating the social transformation of a society, its relationship with economic, social, political, and cultural variables is clearly interactive.

Thus to assert that technology is a dependent variable may simply beg the question. We need to understand more fully the nature of this interactive relationship to be able to determine the relative importance of technology as an instrument for accelerating or indeed resisting social change by reinforcing existing patterns of social stratification, as some observers claim has happened with the so-called Green Revolution in India. Understanding this relative importance and the dynamics of these interactive relationships is crucial if we are to achieve the instrumental goal of being able to design local technology systems and introduce specific technologies to accomplish designated social objectives such as the alleviation of poverty or meeting minimum material needs.

Yet it already is possible to make predictive statements about technological interventions in the alleviation of rural poverty with a fair degree of probable reliability (say three or four times out of five) based on what we now know of the Indian and Chinese experiences. If such interventions are made in an undifferentiated manner into a highly stratified rural community, the principal benefits are likely to flow to the dominant elements in the community, thereby reinforcing or even exacerbating existing inequalities. Only when such interventions are concentrated on the economically and socially disadvantaged segments of the community and not provided to the more dominant elements is there likely to be any significant alleviation of poverty. And even then we do not know whether it is possible to maintain, let alone advance further, the social changes thus achieved in the absence of a more fundamental redistribution of economic and political power in the community.

Local Technology Systems

Comparisons of India and China - Limitations and Possibilities

If reliable predictive statements about the role of technology in the alleviation of poverty are not easily made within one social and historical context, how much more difficult is it to make a meaningful comparison of two very different countries such as India and China. The temptation is strong to make broad and sweeping comparative statements about the two countries because they have both been engaged in trying to grapple with the alleviation of poverty and other urgent economic and social problems over roughly the same time span of the past 30 years.

Yet the two countries started at markedly different points in the late 1940s and have had such markedly different political and economic experiences, with very different kinds of external interactions and relationships that cross-national comparisons are of doubtful validity. Only when we disaggregate by sector or geographical area and differentiate with respect to time is it possible to make meaningful comparisons. Here case-studies of the sort being attempted in this examination of local technology systems and their role in the alleviation of poverty are more likely to be susceptible to comparison, although because they are case-studies they do not lead readily to generalizations about either country, whether in isolation or in comparison with the other.

If such comparisons do have validity, they may also have some instrumental significance. The central difference between two of the cases being studied - Huadong Commune and the programme for small farmers in Gujarat State is that in the former widespread and fundamental economic and political changes largely preceded the development of an institutionalized local technology system while in the latter a first step has been taken in trying to build the capacity to generate technological change in the weaker segments of rural society without making comparable economic and political changes as occurred in China. One of the key questions in the Gujarat situation is whether technology can be a sustained instrument for the alleviation of poverty in the absence of any significant redistribution of economic and political power or whether over time the dominant elements in the communities involved will simply find new ways of reasserting their exploitative influence over the weaker sections of the community. Knowing the answer to this question would tell us much about the limitations to and possibilities for making technological interventions achieve certain social goals and how those interventions should be fashioned or with what other interventions they should be accompanied to achieve a goal such as widespread alleviation of rural poverty in the Third World.

Hopes and Aspirations for Further Research

Deeper study of existing cases over more extended periods of time should help to sharpen comparisons among cases and lead to better understanding of the dynamics of local technology systems. Since local technology systems do not exist in isolation, further examination must be given to their linkages to higher organizational levels and the modern science and technology and industrial sectors. The addition of more cases will help to test the validity of conclusions reached from a more limited number of cases, although valid generalizations on a national basis will remain difficult for both countries, just as will national comparisons between them.

Such results from further research should help us to increase our understanding of the complex relationship of technological and social change. And that increased understanding should strengthen the instrumental significance of the research by enhancing its predictive quality, thereby making possible the design of more effective technological interventions to alleviate poverty and meet the minimum material needs of most if not all of the world's population.

NOTES

1. 'Jig-saw puzzle' scholarship describes the process characteristic of scholarly investigations of contemporary Chinese society in which a picture is gradually constructed by searching through a wide variety of documentary sources and scraps of data from travellers' observations. By contrast in India, the scholar is confronted with a flood of published materials of all kinds and personal observations from a host of different sources; since these data vary greatly in terms of reliability, the task of scholarly investigation becomes one of sifting through a mass of information to find the proverbial needle in the haystack, leading to accurate, balanced insight into a complex social reality.
2. To assert that technology is generally a dependent variable in the process of social change does not mean that it does not interact with and sometimes generate change in social institutions and processes; the relationship is in fact interactive. And there are cases in which technology appears to act as an independent variable, of which the most prominent example is advanced military technology; if it becomes technologically possible for one party in an adversary relationship to build a more destructive weapons system, that party generally will do so in order to increase its advantage over its opponent.
3. Richard P. Suttmeier has developed the concept of 'technological mobilization' in his *Research and Revolution: Science Policy and Societal Change in China*, Lexington, Mass., 1974, pp. 119-147.
4. See, for example, Hans P. Binswanger, Vernon Ruttan *et al., Induced Innovation: Technology, Institutions, and Development*, Baltimore, 1978.
5. Claude Alphonso Alvares, *Homo Faber: Technology and Culture in India, China and the West 1500-1972*, New Delhi, 1980.
6. The preceding paragraphs have been adapted from Carlos Contreras, Charles Edquist, Osita Eze, Ward Morehouse, and Mervyn Wijeratne, *Technological Transformation of Developing Countries*, (Paper based on informal consultations organized by UNCTAD, Geneva, January 1978), Research Policy Program *Discussion Paper Series*, No. 115, Lund: University of Lund, February 1978, pp. 13-14.

THE GROWTH AND DEVELOPMENT OF THE INSTRUMENTS SECTOR IN CHINA AND INDIA - A COMPARISON

Richard Conroy

This article is a summary of part of the research work being done for a doctoral thesis at Sussex University, England. The main part of the thesis is concerned with the detailed study of the growth and development of a science-based industry, the instruments sector, in China, from 1949 to 1972, and involves an analysis of the many issues met in establishing a high-technology industrial sector in a developing country.

The Science Policy Research Unit, Sussex University, had in the early 1970s taken part in an extensive study of the instruments industry in India, and it was felt that if it were possible, a comparison of the respective issues and problems met by each country would be a useful piece of work.

The article is divided into six sections, beginning with a short summary of the size and structure of the instruments industry in developed countries. The problems met in comparing the policies and performance in developing and indigenous instrument-producing capacity in China and India are then discussed. Sections III and IV summarize the initial stages of development in the 1950s, and the subsequent growth through the 1960s in each country. The overall roles of foreign technology and domestically-produced instruments technology are discussed in Section V, whilst the last section summarizes the similarities and differences between each country's developmental experience in this area.

I. THE INSTRUMENTS SECTOR IN DEVELOPED COUNTRIES

In advanced economies, the instruments sector plays a key role in both industry and scientific research. All automated processes, testing, measuring and analytical work in industry are dependent on instrumentation, whilst it provides the tools needed for fundamental and applied scientific research, the quality of which is partly determined by the quality of the instruments used. There is no commonly accepted definition of what constitutes an 'instrument', a problem which makes it very difficult to define this industrial sector with any great precision. However, for the purposes of this section, two major instrument groups used by an OECD report on the sector are examined, 'off-line' or autonomous instruments (analytical, electronic testing and measuring, nuclear, biomedical and electronic and optical microscopes), and 'on-line' or integrated instruments (industrial process control).

Structurally, by the mid-1960s, nearly all industrially advanced countries apportioned over half of their total research and development expenditure devoted to manufacturing industry on a small group of industries. This group has been named the science-based or research-intensive group of industries, and comprises the aircraft, electronics, drugs, electrical machinery, chemical and instruments sectors. A lot of evidence points to the conclusion that innovative activities in these sectors up to the mid-1960s was dominated by one country, the United States.

Production and Markets

The most important instrument producers in the capitalist world are France, Germany, Italy, Japan, Holland, Sweden, Switzerland, the United Kingdom and the United States. These same countries are also the major users of instruments, because of their high technological levels, the size of their research and development systems and their sophisticated industrial system. Users are both private firms and public bodies, and whilst instrument producers enjoy greater freedom when dealing with the former, public procurement often provides manufacturers with a base for starting up production activities. Overall the number of instrument users is growing rapidly, as is the economic and technical importance of instruments in a country's economy.

The instruments sector of most of the above countries is characterized by a small number of large firms and many small ones, with a very large share of the market taken by the large firms. The sector is split between firms which specialize in producing only instruments and those which are diversified and have other product lines than instruments, with the former tending to predominate in terms of output. Structurally, many U.S. firms have either wholly or partly-owned production subsidiaries in Europe and Japan, whilst a few European firms have subsidiaries in various parts of the world, but seldom in the U.S. The manpower employed by the sector usually represents about 0.3%-0.4% of the total manpower of the manufacturing sector, an insignificant proportion compared with other industries. The total production of the countries listed above in 1965 was 2,500 million dollars, of which 1,700 million dollars was accounted for by off-line instruments, with the U.S. accounting for 64%, Japan 8% and European countries 28% of off-line instrument production. Only four countries were net exporters of instruments in 1966. The export/import ratios for off-line instruments in 1966 were: the U.S. 5.7; Germany 2; the U.K. 2; Japan 1.6; all other countries being net importers. The value of the U.S. domestic off-line instruments in the same year was about 60% of the combined markets of all the major instrument-producing countries, representing about 50% of the entire world market. Government procurement has different effects on the industry in different countries, with 40% of domestic demand in the U.S. being directly or indirectly influenced by government funds, whilst in European countries the proportion is around 25%. As far as the technology incorporated into instruments made in the mid-1960s was concerned, there seemed overall to be no sharp disparities between the United States, the major European countries and Japan. In those cases where technical leads in the United States were apparent, it was mostly in instruments related to electronics - electronic test and measuring equipment, modern analytical instruments, nuclear instruments and industrial process control equipment.

Research and Development and Innovation in the Instruments Sector of Developed Countries

With the rapid progress in fundamental and applied research and industrial technology in the last 30 years, a high rate of innovation has been necessary for the instruments industry, which produces tools for both sectors, to keep up and play a key-role in this progress. This high rate of innovation has caused the industry to be highly research intensive, with firms allocating up to 11% of their sales to research and development, with over 90% going to development work. As would be expected there has been a steady increase in the proportion of qualified manpower in research and development to the total manpower employed in the sector.

Expenditure on research and development in the sector in 1964 was higher in the United States than in all the other major instrument producing countries taken together.

One outstanding characteristic of the sector is that in all developed countries, the major part of research and development is performed by the instruments industry itself, although a close and constant dialogue between producer and user seems to be an important component of successful innovations. The extent of public financing of instrument research and development varies from country to country.

A number of factors, in addition to the above, that influence innovation performance, have been identified. Instrument firms in the United States have a higher rate of technical innovation generally than European firms for a number of reasons. The size of the domestic market is an important factor, as the United States market permits economies of scale in the production of new instruments and a spread of research and development costs. The indirect effects of government procurement, coupled with the rapid increase in research and development expenditure in the United States, have both been of importance in the rapid growth of many instrument companies in the United States. This large market has also led to stronger competition in the United States than in other countries, forcing American firms to constantly renew their range of products.

On the other hand, European countries are characterized by small and fragmented markets and a lower level of government procurement with less incentive to innovate. European instrument users are also often reluctant to accept entirely new instruments until they have been proved completely reliable. European instrument manufacturers therefore are sometimes content to keep an instrument in production over a long period, with occasional facelifts, instead of being involved in the costs of completely redesigning it. The receptivity of American instrument users to new instruments is higher and hence this conservatism towards innovation found amongst European instrument producers is less common in the United States. Innovation performance is also affected by the level of liaison between instrument companies and manufacturers of sophisticated components, which tends to be high in the United States where many new components are innovated, and lower in Europe.

Considerable differences seem to exist in the time taken to develop new instruments between the United States and Europe, the former generally taking half of the time of the latter. The reasons for this include the greater concentration of money and manpower on new product developments in the United States, greater capacity for teamwork, and more efficient research and development management techniques.

The role of patents and licences in the instruments sector is unclear, as views on the subject vary widely. Patents are not generally held to provide effective protection, but can provide better bargaining positions in exchanging licences. Licensing appears to play an important role internationally, as many firms not having foreign subsidiaries often grant licences. Overall, whilst other countries ran deficits despite sales to each other and to developing countries, during the 1960s the United States had an increasing surplus from licensing agreements.

II. PROBLEMS IN COLLECTING AND COMPARING DATA ON THE INSTRUMENTS SECTOR IN INDIA AND CHINA

A number of problems arose in both collecting data and comparing the paths of development taken by the two countries in building up an indigenous instrument-producing capacity. The first problem arose because of the essentially different nature of the two projects. The Indian study was conceived as a project to analyse the development of the instruments sector with the aim of facilitating the formulation of further policy proposals concerning the industry's future growth. The Chinese study, on the other hand, attempts to isolate and analyse those issues crucial to Chinese policy-makers in developing a technologically advanced industrial sector, the instruments sector, and estimates the success in implementing the policies adopted.

The basis of such a study, which compares two countries with different political, social and economic systems, rests on the fact that the development policies of both include the general goal of creating a comprehensive, indigenous industrial and research system. The development of a modern domestic instruments sector was seen by both countries to be essential in achieving this goal.

The basis of comparison rests on the similarities and differences of policy proposals, their implementation and effects in each country in their attempts to develop an industrial sector traditionally dominated by a small group of industrially advanced countries. Insight can also be gained on the specific role of technology as a variable in the developmental process by this type of comparative study.

Have China and India, with their different political and economic systems, met similar constraints in developing their respective instrument sectors? Do the technological demands (e.g. quality control, high rates of innovation, specialized production equipment, a comprehensive components industry) of an industry which produces a wide range of sophisticated products incorporating advanced technologies set a certain dependent pattern of development on industrially less advanced countries which in certain respects transcends any one country's ability, through its unique political and economic system, to devise a set of development policies radically different or more 'successful' than any other country at a comparable stage of development? Or are different patterns of growth and development of science-based sectors influenced not so much by their special (e.g. technical) characteristics but more by 'external' factors, be they political or economic in nature or a complex interaction of the two? Other external factors, such as dependence to varying degrees on rapid innovation in a small number of industrially advanced countries brought about by high levels of research expenditure in certain industrial sectors, have also to be considered.

In trying to answer these questions, a number of problems immediately arise. The most important is the variety, reliability and depth of the information available from each country. In both cases, this is compounded by the very fragmentary nature of the industry itself, a difficulty also met when studying the instruments sector in industrially advanced countries. For the Indian study, a number of devices were used to collect information. These included government policy documents, trade figures and interviews with personnel involved in various aspects of the sector's development, all done with the cooperation and assistance of Indian agencies connected with the instruments sector. Combining the information from these sources, a reasonably detailed picture emerged of the sector's history from 1947 to the mid-1960s, both quantitatively and qualitatively. To clarify the size and structure of the industry in the late 1960s, a detailed questionnaire was prepared and sent to all the firms in India that could be identified as producing instruments. This provided a very detailed breakdown of the sector's structure. Interviews were also held with firms that had bought instrument technology from both foreign and internal sources. Although the sample was relatively small, the report was able to isolate certain trends in both types of technology transfer processes. Therefore, although many difficulties were met in collecting the above data, and the conclusions reached were often tentative, the Indian study was able to construct a reasonably comprehensive picture of the instruments sector that enabled policy makers to propose plans for the sector's future development.

The Chinese study, undertaken by the author, was naturally subject to much more fundamental constraints on the amount, variety and reliability of information available. The problems encountered in doing any research work on modern China are well known, and if anything more acute if such a small and relatively specialized subject as the instruments sector is chosen. As so little quantitative data was available, the following approach was adopted. First, an historical analysis of the sector's development was followed, which closely mirrored the general stages of economic development that the Chinese themselves see their recent political and economic development as falling into. Within this historical framework, in each period three sets of policy issues were focused upon. These were the development of an instrument-producing capacity, the growth of the scientific research sector and the acquisition of new instruments technology. In each period, the implementation of these policies was examined, as was the performance of both the instrument-producing sector and the research sector in supplying new instruments technology to the sector. Although there are aspects of the development of the Chinese instruments industry about which very little is known, by using the above analytical and descriptive approach, a number of issues emerge which can be compared with the situation in India. The following sections, therefore, are a brief summary of the initial findings of the comparative study.

III. THE INITIAL STAGES OF DEVELOPING AN INSTRUMENTS SECTOR IN CHINA AND INDIA

Both countries in the late 1940s had a small number of workshops producing very simple instruments and doing elementary assembly and repair work, and therefore neither had a firm basis on which to build a modern instruments industry. In India, the importance of the sector was recognised very early and a report published in 1947 emphasized the following policies that were considered crucial to its successful development. Because of its role in industry and research, it was decided that a future instruments industry should be state-owned, with the government responsible for supplying new instrument technology and training skilled personnel. It was hoped that the industry would move rapidly away from dependence on foreign technology, materials and improvements towards indigenous research and production, and a number of measures were proposed to this effect. In supplying the necessary technology, it was argued that only a broad scientific infrastructure could act as a base for such a science-based industry as the instruments sector. However, by 1956, indigenous production had hardly exceeded early post-indepencence levels, few if any of the report's policy recommendations had been implemented, no new productive capacity had been added, and no organizational or administrative structure had been established to carry out the above policies.

In China, whilst no comprehensive plan to develop the instruments sector appears to have existed before 1957, the First Five Year Plan (1953-7) included plans for building a number of large, modern instruments plants producing a range of instruments and meters for industry, and two such plants were in operation by 1956. During this period, the production of instruments rose steadily, partly due to the contribution of small and medium plants that had been converted to produce simple instruments and partly due to the new modern factories starting up production. These new factories, mostly using foreign (Russian and East European) technology and equipment, were regarded as the core around which the instruments sector would develop. By the end of the First Five Year Plan, therefore, the Chinese government controlled all the means of production of instruments, materials and components, as well as foreign trade and the acquisition of foreign technology. Steps were also taken to organize and develop a comprehensive research infrastructure, with effective links with industry. It is estimated that a self-sufficiency rate of around 60% for instruments was achieved by 1957, whilst imports remained static. This is in contrast with the situation in India, where production remained extremely low, no new modern factories were built and imports rose by about 100%.

By the end of 1956, the problems perceived by Chinese planners in developing the sector were similar to those envisaged by the 1947 Indian report, and included the question of unifying the organization of the industry, coordinating research activities with production needs, changing from repair and assembly to large-scale modern manufacturing methods, as well as increasing the variety and improving the quality of instrument products. Other problems included the lack of specialization, supplying high-quality parts and components, standardization problems, decisions on the location of instrument plants and related research and development and service facilities, repair and maintenance, supply of instruments and questions on the balance between imports and indigenous production. Whilst little was done in India to grapple with these problems, China by the end of the First Plan had invested heavily in building the basis of a modern industrial sector, particularly in machine-building, and there was a rapidly growing demand for all types of instruments. Expertise was also gained in constructing large modern plants.

The early development of an instrument-producing capacity in China was a direct consequence of the policy of giving priority to heavy industry, in contrast to India which invested more in developing its social and agricultural infrastructure.

Reassessment and external change

Towards the end of the 1950s, both countries carried out a reassessment of the policies and plans to develop their respective instruments industries. In India, another report was prepared, and published in 1958. In China, various policy documents concerning the Second Five Year Plan (1958-1962) and the Great Leap Forward (1958-1960) contained a number of policies connected with the development of the instruments sector.

One of the main preoccupations of the 1958 Indian report was to put the industry on a more firm organizational basis. As a result, two agencies were established, one to coordinate production and the other to coordinate work on new instrument technology. Other recommendations concerned the technical side of the instruments sector. Proposals were made to establish a number of regional design and development centres, and standards units at important instrument production areas. The problems that most concerned Indian policy makers were the supply of indigenously produced materials and components, the production of special machinery for instruments and the special problems inherent in supplying the large and diverse range of components needed by the sector. The lack of specialization, and the duplication of capacity and the competition between large and small firms in the production of certain categories of instruments, were all seen as further constraints in the sector's development.

In China the development of the instruments sector in the late 1950s was influenced by five factors. As far as economic and industrial policies were concerned, because of the recognition that the sector was science-based, large-scale capital intensive methods of production were seen as necessary, i.e., the technical aspects were the principal constraining factor. Technically therefore large modern plants were to make up the core with small and medium plants performing secondary functions. The technology utilized at this state was necessarily Soviet and East European, although more domestic research and development work was done in this period. Structurally and organizationally, there was state control over all aspects of the sector. It had its own bureau in the Eight Ministry of Machine Building and its own industrial and research institutes. There was a division of labour between instrument plants and a comprehensive plan had been formulated covering the production, import supply and repair of instruments for research. Politically, the mass mobilization campaigns of the late 1950s does not seem to have had as much relevance to this sector as it had for others, although as part of this movement the education and research systems were encouraged to make more of the equipment they needed themselves. Ideologically, the goal of rapid socialist construction leading to eventual self-sufficiency meant initial dependence on foreign instrument technology to build an independent productive capacity, whilst at the same time increasing the volume of domestic research and development achieved.

Up to 1960 therefore a fairly constant and consistent set of policies had been applied to the development of the instruments sector which whilst formulated within the prevailing ideological and political framework, was heavily influenced much more by factors internal to the instruments sector itself.

In comparing the industrial and technical policies adopted by both countries in the late 1950s, certain similarities stand out. Neither country had been able to generate the technology needed to develop the sector, the major part of which came from various sources abroad. However, both countries reaffirmed the importance of possessing an indigenous instrument-manufacturing capacity, which would contribute substantially to each country's proposed increased tempo of industrialization. Both saw the need to organize the sector so that both research and production could be supervised closely by the state. The importance of large production units was recognized as being crucial to the sector's development, and in both countries a significant proportion of output in some product categories was contributed by a small number of large plants. Certain problems, however, would take a long time to solve, including the indigenous production of materials and components, the low level of specialization, and the production of precision machinery for the sector. The lack of technical services and standards was also seen by both countries to be a serious constraint. However, whereas by the late 1950s, India was just starting seriously to develop the sector, due to external factors outlined below, China was starting to move on to its second stage, that of producing precision instruments utilizing more advanced technologies.

The Performance of the Chinese and Indian Instruments Sectors During the 1950s

Despite the fact that there was little state intervention (although this was government policy), in developing the Indian instruments sector during the 1950s, the performance of the sector improved dramatically from 1956 onwards, especially between 1956 and 1958. The Indian Second Five Year Plan (1956-60) placed much more emphasis on increasing industrial output, and the education sector expanded rapidly, creating increased demand for many types of instruments. However, the economy quickly ran into serious difficulties, due to a rapid depletion of foreign exchange reserves, leading to tight government import controls. This led domestic instrument users to look for indigenously produced instruments, whilst firms trading in instruments started to become manufacturers. The latter did not turn to the domestic research and development sector to supply the needed instrument technology, as at this state it had neither the inclination, organization or expertise necessary. Foreign collaboration was sought as foreign instrument manufacturers could provide one or more elements of a complete package, which included proven designs, machinery, technical and managerial training, a known trademark and products which had been tried and tested in other markets. This initial reliance on foreign technology led to a pattern which was perpetuated throughout the 1960s. The development of the Indian instruments sector in the late 1950s therefore was not primarily due to internal policies and planning, but rather in response to outside economic pressures.

In China, available evidence points to rapidly increasing output during the late 1950s as a number of large modern plants started production, and a doubling of the workforce in the period of a few years. Apart from institutes specializing in instruments research, the instruments sector had three potential sources of instruments technology to call upon: the Academy, higher education and ministry research sectors. Although it is difficult to gauge with any precision the relative proportion of new instrument technology they contributed to the production sector, all did so to a certain extent. The Academy sector tended to develop more sophisticated instruments for research work, whilst the higher education sector developed and built instruments in order to equip itself. The ministry sector was more concerned with the modification of foreign designs, and appears to have been the sector most involved with the instrument-manufacturing sector. Whilst, however, the instrument sector *started* to obtain some new instrument technology from a variety of domestic sources during this time, most of the technology needed further to develop the sector was acquired from abroad. This was done through trade and science and technology cooperation agreements with the Soviet Union and East European countries. In the case of China, therefore, complete state control was exercised over the type and scale of instruments technology imported (though the choice was limited to what was offered) in contrast to India where the technology was imported by manufacturers in the private sector with little government control at this stage.

The performance of the Indian instruments sector in the late 1950s improved because of unforeseen and totally unplanned-for factors, and these factors set the general pattern of the sector's development for the next decade. The sector's rapid growth in China can be attributed to two factors. Complete state control after 1956 over all aspects of industrial planning, foreign trade and research policy facilitated integrated policy decision for developing the instruments sector. The rapid increase in output, product variety and precision was a direct consequence of the reasonably successful implementation of a number of technical and industrial policies specifically designed to stimulate the sector's growth.

IV: THE GROWTH AND DEVELOPMENT OF THE CHINESE AND INDIAN INSTRUMENTS SECTORS DURING THE 1960s

In India, the instruments sector continued to grow during the 1960s most of it concentrated in the private sector and using foreign technology, as efforts to use the Indian research sector as a source of instruments technology met with relatively little success. The Sino-Soviet dispute had far-reaching repercussions on all sectors of the Chinese economy, especially in the heavy and science-based sectors. This, coupled with a basic reorientation of development priorities, acted as a constraint which forced China largely back on its own resources, as well as necessitating a search for new foreign sources of instrument technology. The Cultural Revolution further modified industrial and research policies that had been in force during the early and mid-1960s, and in some respects was a stimulus to the instrument sector's development.

The Indian study has identified a number of factors that determined the structure and size of the instruments sector during the 1960s. The general industrial expansion of around 60% between 1960-5 created a demand for various types of instruments, especially those used in industry, and there was a rapid growth in domestic production in this area. This was similar to the Chinese experience, as one of the major aims of the instruments sector during the 1960s was to provide equipment for large modern plants, refineries and power stations, without outside aid. The Indian government also took a much more active role in the sector's development during this period both in terms of control and planning. The agencies set up to co-ordinate production and research, however, tended to operate in ways not originally envisaged, efforts in both areas being dissipated and rather ineffectual, with the result that no one agency in either area was in complete control. Overall, though, by 1968 the instruments sector had produced a wide range of products and satisfied around 70% of the national demand, i.e. India was fairly self-sufficient in terms of output but not in terms of technology generation. An analysis of the sector's structure and size in 1968 reveals a number of similarities with the Chinese situation. In India, around 70% of the instrument firms employed less than 40 workers, whilst only four employed over 600. Whilst there were many more large plants in China, it is likely that the majority of Chinese instrument plants were also small and, like the Indian case, poorly equipped. Total employment in the sector in India doubled between 1963 and 1968, reaching 23,000. This compares with a total of 70,000 in China in 1959, and whilst it is difficult to draw conclusions from such figures, it is obvious that by the end of the 1960s, the Chinese instruments sector was many times larger than the Indian sector. The Indian small-scale sector, as would be expected, produced only simple types of instruments, and was often in competition with larger firms making the same products. In China, it was not until after the Cultural Revolution that some small specialized plants were set up, making high precision instruments. Another difference is that China used its production resources better than India, by following the policy of larger plants concentrating on new and more sophisticated products, whilst small plants took over the production of the more simple instruments.

The Indian sector showed a highly-skewed structure in terms of output, in that the 50 largest firms, 7% of the total, between them produced around two-thirds of the total output. Whilst a similar situation occurred in China in the late 1950s, the policy in the 1960s of building small and medium plants in all sectors would have corrected this tendency somewhat. In India, the product variety did change over time, away from those traditionally produced (e.g. survey, laboratory and education instruments) towards industrial and research instruments. This was also true in China, although the transition took place earlier and was much wider in scope, with emphasis during the 1960s on the development of industrial process-control analytical instruments and advanced research instruments, with progressively more sophisticated models being trial-produced throughout the decade.

In India, a few large firms produced a significant proportion of the total output in some product categories, a situation that had prevailed in China in the late 1950s, but which was less evident ten years later. Certain areas, however, in China, did tend to specialize in the production of specific groups of instruments. By 1969, there were still many gaps in the types of instruments that India could produce. This was much less apparent for China, partly because of the increasing amount of research on instruments and the technologies associated with instrumentation (e.g. automation, optics, electronics). This in turn was in response to the difficulty of access to advanced foreign instruments technology, much of which is crucial to modern military technology.

The Indian study found no evidence that a rational regional policy had been followed in the 1960s, in marked contrast to China, where a policy of building up an instrument-producing capacity in each economic region and new industrial bases in the interior is clearly discernible throughout the 1960s. The diffusion of production capacity, however, did not generally extend to the trial production of highly complex instrumentation, as this activity was usually carried out in the old, highly industrialized centres where instrument production had originally been concentrated.

The Indian study identified a number of factors which were seen as constraints on the development of the sector into the 1970s. The inability to produce domestically a range of parts and components was one of these, compounded by the difficulty in being granted licences to import components that could not be produced in India. Although China, by the late 1960s, still had to import some specialized components, great attention was paid to developing a comprehensive components sector, and such branches as precision machine-tools. This development was partly a consequence of the decision in the late 1950s to set up research units working on a number of advanced technologies, and partly because of the problems involved in procuring specialized components from abroad. The other main problems seen from the view of Indian instrument manufacturers were connected with specific government policies towards the sector and included government procurement procedures. Issues that were *not* regarded as important by Indian manufacturers included questions of quality control, product improvement and the training of scientific and technical personnel to work in the instrument sector. The Indian study regarded this as a reflection of a protected market and a rapid increase in demand.

In contrast, the most enduring themes in Chinese industrial policy documents throughout the 1960s *vis-à-vis* the instruments sector and industry as a whole, were connected with such problems as raising the quality of products, increasing variety, expanding indigenous design capabilities, introducing new technologies into industry (e.g. automation, electronics etc.) and making research more closely meet the needs of industry. Although none of the problems were completely solved, a variety of measures were taken to alleviate them.

To sum up, in the Indian case there is no evidence that new policies were introduced specifically to develop the instruments industry during the 1960s, rather some previously formulated policies were implemented, not very successfully, to try to control and plan the sector's growth. The development that did occur depended mainly on access to western technology, on a steadily growing domestic demand, tight import controls and a heavily protected home market. Government policies formulated to develop industry in general, which were themselves partly influenced by external factors, had much more impact than those specifically devised to develop the instruments sector.

The biggest factor influencing the development of the Chinese instruments sector during the 1960s was ideological - the Sino-Soviet rift. This abruptly changed the tempo and direction of development as a whole. Much more rapid self-sufficiency was seen as a prerequisite for national survival. For a number of reasons, China, unlike India, could not buy the instruments technology it needed directly from the West. Whilst a number of industrial and technical policies that affected the sector continued to be re-emphasized and implemented, some new ones appeared, designed to meet the new conditions e.g. emphasis on small and medium plants, concentration on instrumentation connected with defence, a new regional location policy. There is no evidence to show that the sector was significantly disrupted by the events of the Cultural Revolution, apart from developmental work. Indeed a number of policies emerged that could have had a beneficial effect on the sector's development, except in the area of basic research instrumentation.

As far as performance is concerned, both countries made considerable progress in terms of output and increases in the variety of products manufactured. However, because of the early development of the sector in China, despite the setbacks of the early 1960s, the vigorous self-reliance campaign resulted in a significant expansion not only of production and variety but also innovative activities. The structure of the sector in India in the late 1960s developed along similar lines to that of China in the late 1950s. The latter had, by this time, moved on to a new stage of development, innovating and producing a range of more complex instruments for both industry and research, as well as supplying many of the special materials, components and machinery needed. A cursory glance at the development of advanced fields either using instruments or supplying components to the sector (e.g. missiles and nuclear weapons, space research, computers, electronics etc.) in the two countries in the late 1960s shows that China was ahead significantly in terms of products and indigenous development.

Richard Conroy

V: THE ROLE OF FOREIGN AND DOMESTIC TECHNOLOGY IN THE GROWTH OF THE CHINESE AND INDIAN INSTRUMENT SECTORS

Foreign Technology

There is no doubt that the Indian instruments industry was up to the late 1960s heavily dependent on foreign technology, as over 2,400 such agreements were signed between 1956 and 1965. Industrial instruments formed by far the most important category, and firms from the United Kingdom, Japan and West Germany accounted for 70% of all agreements. These three countries were also China's main instrument trading countries during the 1960s. The main elements of technology transferred were the brand name, the design and the production techniques, although the majority of Indian firms who were party to such agreements claimed to have made some modifications in the technology transferred. The average time that elapsed between the initiation of talks and the start of production was rather long, about 3½ years.

China was largely dependent on Soviet and East European instrument technology in the 1950s both in embodied and disembodied form (e.g. complete instrument plants, special machinery, designs, production techniques, blueprints, prototypes, technical information, advice and training) and a continued role was seen for such technology in the 1960s. The Sino-Soviet dispute effectively cut China off from its traditional sources of technology. As was the case in India, pressures unconnected with the development of the instruments sector *per se* forced new policies to be adopted. However, the Chinese response was to lessen the intended role of foreign technology and speed up the policy of self-reliance in supplying internally the required instruments technology. A policy of seeking new technology from the West was also pursued, but the elements of technology transferred were of a fundamentally different nature from those transferred to India, and its contribution to the development of the Chinese instruments sector was very modest. This was partly due to the reluctance of Western countries to sell to China products which contained elements of advanced technology. China therefore imported Western products which it urgently needed but could not yet manufacture, i.e. it did not import 'technology' as such. However it also appears to have applied 'reverse engineering' to some Western instrumentation as well as modifying the design of Soviet instruments bought in the 1950s.

Growth of the Instruments Sector

The Indian study concludes that overall most of the agreements it looked at in detail could be described as being to various degrees successful. This was mainly due to the fact that the foreign supplier, being a commercial unit, could transfer a whole range of skills and equipment necessary to set up and maintain the manufacture of a tried and tested product. The Chinese response to acquiring foreign technology was much more diffuse than the Indian response, in that no agreements were signed to produce foreign instruments under licence. Imports of instruments from the Soviet Union fell dramatically and all technical cooperation ceased. Between 1961 and 1964, trade with Western countries in instruments slightly exceeded that with the Soviet Union and East European countries. From 1964 onwards, trade with Western countries increased drastically, with sales being fairly evenly divided between industrial and research instruments. The Chinese also maintained a lively interest in keeping up with the latest developments in a number of high technology areas including instrumentation. Despite the hard-line attitude displayed over importing foreign technology during the Cultural Revolution, technical information on the latest foreign instruments was still sought. To what extent, however, imported Western instruments were copied is an extremely difficult question to answer. Unlike the Indian case, Chinese scientists and technicians, in copying a foreign product, would still have had to do a large amount of development work, especially for more complex types of instruments which demanded the possession of a wide range of design, development and production skills. Overall, the choice was made to use foreign instruments technology as a 'back-up' force in developing China's instrument sector during the 1960s, with the main emphasis on indigenously generating the required technology, and consequently the role of foreign technology was fundamentally different in China compared with India.

Internal Technology Transfers in China and India

The most striking point to emerge from the Indian study's work on internal technology transfers was the small number that had taken place during the 1960s. A large proportion of these were judged to be only marginal successes due to a number of problems. This was partly because of the type of technology involved, which tended to be more complex than that involved in foreign technology transfers. The main problem centred around the inability of the Indian research units involved in transfers to provide the facilities and expertise needed to turn a laboratory prototype into a reliable production model. This is not really surprising, as although the research sector had originally been set up to provide Indian industry with an indigenous technical base, its activities became oriented much more towards research which had little immediate economic or industrial use. This happened despite repeated government injunctions calling for the research sector to be more responsive to the needs of industry. Attempts to establish research agencies whose specific role was to carry out industrial research have also not met with marked success. The advantages seen in using domestically generated technology were that it cost less, the product could be more easily adapted to meet local conditions and resources, and the often long delays in obtaining government approval to import foreign technology would be avoided. It was felt that one of the reasons why many instruments that had been indigenously innovated by Indian research units had not been transferred to the production sector was because of an information gap between the two sectors. However, most of these unexploited instruments were developed by the units concerned for their own use, and would not have had a very wide commercial market. This, though, together with the fact that a number of research units developed their own instruments and in some cases sold them to other laboratories, meant that the capacity to develop instruments domestically did exist in the Indian research sector, but that the mechanisms whereby this capacity was put to wider use were faulty i.e. the problem was structural rather than technical.

In China, the problems of internally generating instruments technology were of the greatest importance. The role of the three major sources of new technology, the Academy, education and ministry sectors, changed during the 1960s as attempts were made to make them more responsive to industrial needs. After the Cultural Revolution, production units themselves also did more development work, and an informal industrial research sector was also set up.

Up to the Cultural Revolution, the Academy and education sectors can be most closely compared with the Indian research sector, as the former developed a range of instruments for their own use and also cooperated with industry in developing new instruments that were put into production. It is clear, however, that both the Academy and to a lesser extent the education sector, were much more responsive to the needs of industry than the Indian research sector. The ministry research units to a certain extent paralleled the role of foreign instrument firms in India, and played an increasingly dominant role in the mid-1960s. It must be emphasized that development of instrumentation was by no means confined to those units specifically set up to do so in each research sector, but was carried out by a wide range of research and development units. It appears that in neither country did the instrument-manufacturing sector itself directly generate new instruments technology, unlike the situation in industrially advanced countries.

These very generalized similarities, though, tended to break down after the Cultural Revolution. The role of the ministry sectors appears to have remained unchanged, whilst the other two sectors reoriented their research activities much more towards supplying industry with new technology, and instruments technology figures in this new list of priorities. The research sector itself was also more active in building its own equipment. Factories were encouraged to take more part in research and development work as well as to build the machinery and equipment they needed (e.g. industrial process-control equipment) and this work was carried out by established engineers and informally trained worker/ technicians.

Although China and India both set up their research sectors in order to contribute to their economic development, the continued insistence in Chinese development plans, economic and industrial policies and ideological campaigns that the activities of the research sector be integrated with the economic and industrial development, undoubtedly made it more effective as far as the instruments sector was concerned. However, even though the state exercised more control over the content of research, by the mid-1960s it was found to be necessary both to reorient certain parts of the research system to make the work more appropriate to Chinese needs, and also to break down the barriers that prevented new technology from being transferred to industry (e.g. by sending scientists into factories, forming special technical innovation groups in plants, training worker/ technicians, increasing ties between research units and factories).

These two problems, that of research activities not being appropriate and the failure to scale what the Indian study calls the technology/production interface even where technology is relevant, are the major reasons why India failed to generate internally the technology needed by its instruments sector, tasks in which China was significantly more successful, though not without many problems and difficulties.

VI. CONCLUSIONS

It is concluded that the nature and characteristics of the instruments industry impose certain significant constraints on any developing country in developing the sector. Both China and India confronted a number of similar problems in developing the sector, and many of the proposed solutions were similar. These centred around initial reliance on foreign technology and close state supervision or control over the organization and activities of the research into and production of instruments. China's much greater success in developing a fairly independent instrument-producing capacity can be attributed to its political and economic system allowing greater scope in both formulating and implementing specific policies and plans to establish and develop such a science-based industry as the instruments sector and those sectors allied to it, which were integrated with wider industrial, economic and developmental policies.

However, the rate and direction of development of the instruments sector in each country, in the long run, was influenced more by wider political and economic considerations (and in China's case by ideological factors) that had nothing to do with the sector *per se*, than by narrow technical and industrial policies. The concept of self-reliance applied to this sector was certainly interpreted differently by each country, and implemented much more successfully by China.

SURVEY OF INDIA-CHINA COMPARATIVE RESEARCH

Erik Baark

This chapter presents a survey of comparative studies on India and China, arranged alphabetically under the following headings:

A. Economic Performance
B. Development Strategies and Ideology
C. Rural Development and Agricultural Production
D. Education and Science
E. Technology, Health and Foreign Trade

Each group is introduced by a short note, where general comments concerning the subject, scope, methodology, etc. of items covered are presented. The studies are then summarized and their contribution to India-China comparative research is evaluated with reference to results and methodology. The description of each item is not intended as a simple annotation of the study, but as a critical assessment of contents and methods applied.

Full bibliographic information for the items can be found in the Bibliography of India-China Comparisons at the end of this volume.

A. Economic Performance

One of the major fields in comparative research is concerned with the evaluation of the economic performance of various countries. Economists have been comparing the achievements of developing countries along several lines, where the growth rates in gross national product and per capita consumption have received special attention. In particular, economists from the United States and from India have shown great interest in comparisons of the economic performance of India and China.

It appears that apart from an interest in the economies of large developing countries, the objectives of these comparisons have often been to show the superiority of particular ideologies associated with one or other of the countries. The problem has frequently been formulated in the following manner: if China/ India has achieved a superior growth rate in the gross national product, this can be interpreted as a superiority of socialism/liberalism respectively. The issue is obviously problematic, and as a consequence the conclusions reached by various researchers seem to have been significantly influenced by the political implications of their analyses. This becomes evident when the papers by the conservative Indian scholar Subramanian Swamy (items 4 and 5) are compared with those of the American scholar Thomas Weisskopf (items 6 and 7). Dealing essentially with the same period, from national liberation to the 1970s, these two scholars manage to present totally contradictory pictures of the economic development achieved in the two countries.

One reason why two studies of the same period can reach such different conclusions may be found in the fact that figures for both the size of the population and the national product of China have been based on estimates. These estimates are subject to frequent revisions, and high or low figures may be chosen almost at will. Such figures are, however, essential for the calculation of the rate of growth in gross national product per capita. Thus, when for example value-added figures for gross national product - the weighing of domestic product in prices being a process which may also yield entirely different results - are combined with various figures for the size of the population, it appears to be possible to reach almost any result.

A general conclusion reached after reading the items covering economic performance is that such studies must be classified as normative-descriptive. Though studies of economic performance seem to be important in India-China comparisons, they appear to have limited relevance for the subject of this volume: technology and science for development.

1. Debesh Bhattacharya: 'India and China - Contrast and Comparison 1950-72' (1974).

In this paper the achievements of the Chinese economy are compared with those of India for the period 1950-72. A postscript describes comparative growth in food production during the period 1969-73, rates of growth in industrial production for 1973 and 1974, and relative developments in income distribution, unemployment, and food consumption.

In section 2 of the basic study statistics are presented on rates of growth in grain output, industrial production and national product. These are based on official Indian sources and as many as eleven different estimates (in the case of national product) for China. Of particular interest are a table showing changes in the rate of growth of selected items during five base-year periods, and a table comparing Indian and Chinese rates of growth with those of advanced countries in periods of similar development.

Bhattacharya concludes that 'China's economic growth record appears to be substantially better than India's to date'. It is argued that the achievements of the Chinese in issues like unemployment problems and economic infrastructure further emphasizes the success of the Chinese model.

2. Sarwar Lateef: Economic Growth in China and India 1950-1980. (1976).

This report from the Economic Intelligence Unit is the most recent comprehensive quantitative comparison of economic performance in India and China. It presents a vast amount of statistical material, and - in addition to a discussion of the data employed - it incorporates a critical overview of the policies pursued in the two countries.

Economic growth in China and India during the period 1950-75 are described in separate chapters. The description also includes an analysis of the development in foreign trade of the two countries. The conclusion indicates that China's growth rate has become increasingly higher than India's during the period 1960-1974, but - in contrast with India - has fluctuated considerably. In addition, the comparatively higher growth rate of China is entirely due to rapid industrial growth, while the rate of growth in agriculture has been larger in India.

In an interesting discussion of the comparative prospects of each nation, Lateef presents projections of high and low growth rates and analyses the factors which may influence the comparative economic growth of the two countries. For example, the author argues that a major problem for China is the present high level of yields per hectare, which makes scientific and technological progress the most pressing task. India, on the other hand, has a problem of creating the preconditions for increasing yields which are nowhere near the technological ceiling.

3. K.R.M. Rao: 'India measured against China'. (1976).

This paper is a short note reviewing the economic performance of India and China, unfortunately without documentation of the arguments presented. The main part of the paper is devoted to a discussion of the inadequacy of conventional statistical methods in comparisons of developing economies.

The Chinese model of development is considered much more appropriate to a developing economy, and it is argued that 'if the average per capita income of the poorest 5 to 10 per cent of the population is the criterion, China seems to have done better than most countries'.

4. Subramanian Swamy: Economic Growth in China and India: A Comparative Appraisal. (1973).

This paper by Subramanian Swamy deals with economic growth in China and India during the period 1952-70. In his analysis, Swamy explicitly focuses on a quantitative measurement of the economic performance of the two countries, whilst the social sector is treated as a quantitative performance of the 'residual' sector, rather than in relation to such issues as social development. In addition, the objective of comparing the rate of growth in national income means that the output of the agricultural, industrial and service sectors must be recalculated to yield a 'net value-added' figure.

In this process, Swamy makes several assumptions which (deliberately?) lead to a downgrading of the Chinese performance. For instance, the value of output in each country is calculated in 'parity rupees' - i.e. 1948-9 Indian prices; and the service sector in China is assumed to have experienced a sharp decline during the Cultural Revolution.

As a consequence, the conclusions reached in the analysis are rather controversial: '(a) the growth rate of the Chinese and Indian economies has been about the same over the period 1952-70, and (b) there has been a significant deceleration in growth rates in China, especially during 1952-65, and a significant acceleration in India'.

5. Subramanian Swamy: 'The Economic Distance Between China and India, 1955-73'. (1977).

This paper compares the physical output of selected items in selected base-years from 1955 to 1973. By calculating the ratio among outputs of such items as foodgrains, cotton cloth, sewing machines and crude oil for India and China in these base-years, Swamy argues against the belief that China's output is larger for every major commodity. Thus, the economic distance, i.e. difference in output, has not increased between 1955 and 1969, but since then China has been getting increasingly ahead of India.

The methodology of comparing total or per capita output of selected items again underscores the biased view of development held by many economists. The 'economic distance' measured through this method appear to have little relevance in the context of developing economies, and the methodology does not appear to have gained much popularity.

6. Thomas E. Weisskopf: 'China and India. A Comparative Survey of Performance in Economic Development'. (1975).

This paper contains an analysis of the economic development of India and China in the period 1950-1975, but concentrates on the most recent situation in the two economies. While a description of differences in rates of growth, sectorial development, etc. more or less similar to other analyses of economic performance is presented, most of the paper is devoted to a description of achievements in each nation with respect to economic equity and economic self-reliance. In the description of economic equity, selected indicators such as income distribution and urban-rural differential are used. Indicators such as national control, foreign debt, and technological independence are used for economic self-reliance.

The author concludes that 'the evidence compiled and presented in this paper leaves no doubt that China's progress in three important dimensions of economic development has been significantly greater than India's in the modern post-War period'. This conclusion appears to be premature, considering, for instance, the results reached by researchers comparing agricultural development. Still more questionable is the value of some of the evidence indicating achievements, such as visitors' reports and impressionistic evidence (*sic!*), and some implicit normative assumptions concerning good or bad performance, for example, that technological independence is good on all accounts. Furthermore the choice of indicators tends to carry a bias in favour of areas to which China has been devoted more attention than India. The point is not that China's achievements in self-reliance are less than India's, but that the normative approach applied throughout the paper is not explicitly stated.

7. Thomas E. Weisskopf: 'China and India: Contrasting Experiences in Economic Development'. (1975).

This paper appears to be a revised version of the preceding item. Much of the statistical material presented in the preceding item has been omitted in this paper, but it contains a longer discussion of approaches to economic development and possible sources of differential progress.

The author argues that a number of elements in the Chinese approach have been conducive to economic and technological progress, for instance, the radical transformation of social and economic institutions and greater egalitarianism. In addition, the strong central government of China is able to implement its policies with far greater effectiveness than the Indian leadership. This, the author argues, is the result of an approach which has succeeded in shifting the political power away from those classes which benefit from inequality and foreign involvement in the economy. A similar process does not appear to have occurred in India.

B. Development Strategies and Ideology

Comparisons of development strategies and development models are quite frequently encountered in India-China comparative research. The reason for this is obvious: the experience of India and China as large developing countries may present guidelines for development policies in other countries. In addition, the contrast between socialist and liberal ideologies and their impact on the development models adopted by the two nations has attracted considerable interest among scholars.

The development models can, however, be studied in various conceptual frameworks and from different points of view. Thus Anderson explores the differences and similarities in the ideologies of leading political figures in an attempt to trace the roots of self-reliant development strategies. Bergman describes the transformations of agriculture in their socio-political context, while Hettne is more concerned with the evolution of political strategies for economic development and their social and historical roots. Finally Byres and Nolan are basically concerned with the respective approaches of India and China to a particular aspect of development strategy: social and economic inequality.

All the studies reviewed in this group are descriptive presentations of the two development strategies or aspects of these. It was argued in Chapter I that descriptive studies are valuable as background material for analytic studies, but that they do not present theoretical explanations as such. The survey of studies in this field appear to confirm this viewpoint.

A special variety of India-China comparisons is the description of the political ideologies of Mao and Gandhi. On the assumption that the political viewpoints of these two statesmen were products of their respective cultural heritage, and that they in turn significantly influenced the development of their countries, such studies might bring out important comparisons of cultural contexts of development. While the studies by Anderson and Vaitheswaran present steps toward such comparative insights, the paper by Sethi shows, on the other hand, how easily such studies may become personal crusades against one or another of the ideologies involved.

8. Mary B. Anderson: 'Self-reliant Development: A Comparison of the Economic Development Strategies of Mohandas Gandhi, Mao Tse Tung and Julius Nyerere'. (1978).

Encompassing a comparison between the political thoughts of the three political leaders associated with the self-reliant development strategy, this thesis describes the formation of economic theories in the Western world, the principles for self-reliant development formulated by Gandhi, Mao and Nyerere, and finally the economic development achieved with this strategy in China and Tanzania.

In her treatment of the principles proposed by the three political leaders for self-reliant development, Anderson focuses on planning, collectivisation, consumption and other economic concepts. Of particular interest to the theme of the present volume is a section on the role and use of technology. The author argues that there is a basic similarity in the ideologies of Gandhi, Mao and Nyerere. This results in an approach to technology - the self-reliant strategy - which imposes values for the choice of technologies other than those dictated by the industrialized countries. The basic elements of this strategy are an approach to production which keeps work at the centre and an approach to consumption which subordinates material affluence to other values consistent with social goals.

Finally, Anderson concludes that the self-reliant strategy does present a serious alternative to the Western approach to development.

9. Theodor Bergmann: The Development Models of India, the Soviet Union and China. A Comparative Analysis. (1977).

This volume includes a description of the three models of development exemplified by strategies pursued by India, the Soviet Union and China. The comparison is made through parallel descriptions of the three countries with respect to socio-political and economic developments. The focus in the first three sections of the book is on agricultural development and its effects on the national economy, though performances in industry and education are equally discussed in relation to economic development.

In the last section an integrated comparative analysis is presented. This analysis discusses key concepts in development such as the impact of agrarian reform, unemployment and choice of technology, and compares Indian, Soviet and Chinese approaches to the political issues associated with these concepts. In addition, some of the problems involved in quantitative comparisons of economic achievements are discussed.

Bergmann considers and evaluates development problems from the point of view of a critical, independent Marxism. As a consequence, his treatment is particularly strong in its discussion of social transformations and the approaches to socialism adopted in each country. The text is also provided with tables and diagrams illustrating quantitative indicators of economic and social developments.

10. T.J. Byres and Peter Nolan: Inequality: India and China Compared 1950-70. (1976).

This volume was issued by the Open University as part of a course on patterns of inequality. The essay is consequently basically concerned with approaches to inequality in India and China.

Before treating the Indian and Chinese approaches to inequality, the paper describes the economic strategies pursued and the economic performance attained. It is argued that China, besides outpacing India in the development of industry, has acquired a better position for future economic development through her economic strategy. This strategy has stressed development of producer-goods industries and minimized dependence on foreign capital.

The analysis of inequality in India and China starts with a broad description of the policies propagated in each country with respect to problems of inequality. These policies, which indicate commitments to the solution of the problems in both countries, appear to have met with contrasting conditions at the level of implementation, where China has been more successful in alleviating inequalities than India. The authors argue that the explanation for this contrast must be found in the different politico-economic structures involved in each country, in particular with reference to the 'modes of production' and class structures.

A considerable amount of material is presented to illustrate the dimensions of inequality, which are described in terms of four socio-economic aspects: inequality between regions, urban-rural inequality, and inequalities within urban and rural units. Analysed explicitly from the theoretical view of political economy the treatment focuses on the relationship between development strategies and class structures.

11. Kuan-I Chen and J.S. Uppal (Eds.): Comparative Development of India and China. (1971).

This volume presents scholarly writings from specialists on India and China. The material is arranged under the following headings:

I. Emerging Patterns in Economic Development
II. Development Policy and Planning Techniques
III. Demographic Patterns and Population Policy
IV. Capital Formation and Development of Human Resources
V. International Economic Relations - Trade and Aid
VI. Looking Forward

The material is definitely outdated by now, since most of the papers were originally published in the 1960s. While the idea of selecting articles by specialists in each field is laudable because it can bring relevant material together in one volume, this sort of publication suffers from the absence of any comparative conclusions as such.

12. Nigel Harris: India-China. Underdevelopment and Revolution. (1974).

This item appears to be a collection of essays originally published in 'International Socialism'. Written on the basis of a Marxist viewpoint, the book comprises the following parts: (1) India and the Crisis in South Asia, (2) China and the Struggle for National Liberation, (3) Issues in Marxist Theory, and (4) Economic Development and the Perspectives for Revolution.

The first three parts are mainly concerned with some key issues in Marxist theory such as the revolutionary role of the peasantry, the class structure of the Chinese revolution, etc. Historical developments are described, but the main emphasis is on recent political developments, notably the Cultural Revolution in China and the period of Emergency in India.

In the final section, the economic performance is briefly reviewed. The comparative description is not particularly Marxist and appears to be based on sources from the United States Central Intelligence Agency. At no point does the book provide more than superficial comparative conclusions, and as a Marxist study of development models, the treatment suffers from a lack of explicit theoretical framework.

13. Björn Hettne: Utviklingsstrategier i Kina och Indien. (Development strategies in China and India). (1979).

This book is a comprehensive treatment of the historical background and modern evolution of development strategies in India and China.

Hettne identifies three important social factors in the development process: the power structure, the development strategy, and the economic heritage. The development process is then analysed in terms of the historical conditions of these factors and their dynamic interplay over time. In the first part of the book, the historical roots of the Indian and Chinese societies and political conditions are described. The description deals with, for instance, the influence of imperialism in India and China, the question of nationalism, and the attitude of the peasants toward revolution. In the second part of the book, the dynamic relations between power structure, social developments and development strategy is described for each country. This description focuses on economic and political developments as they are reflected in the formulation and implementation of five-year plans during the period 1950-70, while the developments of the 1970s are analysed with reference to the fate of populist strategies.

A conclusion summarizes the evolution of Chinese and Indian strategies, the relevance of the experience of these countries to other developing countries, and the future prospects for populist development strategies in India and China.

14. J.T. Macrae: 'India and China - A Comparative View for the 1970s'. (1971).

This article reviews three common elements in overall development strategies in India and China, namely

(1) the limitations posed by population growth and the present structure of the population, (2) the development of science and technology, and (3) the progress of the policy of yield-raising in agriculture.

Although the study is based on data from the 1950s and 1960s, it presents projections for the 1970s as well. While these projections may seem a little outdated by now, the generalized comments made on the problems faced by the two nations appear to be relevant even today. The prospects for development are considered mainly in terms of the increase and composition of the populations of India and China in the future. These prospects, it is argued, are clouded by the issue of birth control.

15. J.D. Sethi: 'Mao and Gandhi: Convergence and Divergence - A Preliminary Note'. (1977).

Sethi's paper examines the similarities and contrasts between the philosophies which have come to be known as 'Maoism' and 'Gandhism'. In addition to an exposition of the two systems of ideas and practice, the paper also includes a description of the historical role of Mao and Gandhi in the world - and their attitudes toward this world.

The comparison, however, includes many personal judgements which are in a definitely anti-communist vein. For instance, Marxist 'concealment and distortion of truth' is contrasted with Gandhi's 'search for truth', communist 'dictatorship and elitism' is contrasted with Gandhian principles of 'non-violence' and 'distribution of political power', etc. In terms of analytic insight into these philosophies, the essay seems to have very little to offer.

16. R. Vaitheswaran: 'Gandhi and Mao: A Comparison in Terms of Relevance for the Politics of National Liberation and Reconstruction'. (1976).

This article is concerned with the politics of liberation from imperialism as well as the politics of development. It studies some of the more important similarities of the Gandhian and Maoist ideologies and practice, and is thus intended to cast light on the ideas and circumstances which influenced the ideologies of the two statesmen, who in turn came to influence immensely the political development of their respective countries.

Vaitheswaran examines the impact of the West on both statesmen, and describes for instance the influence of Christianity and British concepts of liberty on Gandhi's ideas of non-violence and equality. He also traces different aspects of Mao's Marxism from anarchist, Leninist, and classical Chinese thought. Other comparisons are made of Mao and Gandhi in relation to populism, social visions, and modernization.

A strong strain of nationalism is detected in the formation of the thought systems of Gandhi and Mao, and the success of both as nation builders are seen as a consequence of their roots in national philosophies; at the same time their success (or failure) in modernization is also a reflection of the contradictions between foreign and indigenous ethical values.

C. Rural Development and Agricultural Production

The papers surveyed in this section can be divided into two groups: studies concerned with economic performance in the agricultural sector and studies devoted to particular aspects of rural development.

The studies on agricultural production generally focus on two subjects: the growth rates achieved in production of food grains, and the social changes which have taken place in the rural sector. When comparing the growth rates, these studies are subject to the same reservations on the methodological issues that were mentioned in the section above on studies of economic performance. Most of these studies, e.g. Bandyopadhyaya and Bardhan, are conducted within the framework of an input-output analysis, that is, output of food grain considered in relation to industrial and manpower inputs. Such analysis suggests that India has achieved more output for the same inputs as China; collectivisation is invariably blamed for China's failure. From a methodological point of view, however, input-output analysis may not be justified in the comparison of agricultural economies which are so different from the outset, for instance with reference to yield per hectare.

Another approach is used by Etienne in his discussion of prospects for food grain production in India and China. In this paper the main focus is on the present capacity for food grain production and the chances for future improvements through organizational change and increased inputs.

Finally, the studies on rural development by Jain and Sigurdson are primarily devoted to an analysis of the effects of political and social organization on the rural economy.

17. Kalyani Bandopadhyaya: Agricultural Development in China and India. A Comparative Study. (1976).

This study examines the historical development of the agrarian structure from the 1930s to the 1970s, describing, for instance, the land reforms and the establishment of the rural People's Communes in China and Community Development projects in India. Finally, the agricultural production of China and India are compared in quantitative terms.

Although the author emphasises the difficulties of an input-output analysis of the development of agricultural production in India and China, the concluding remarks are nevertheless kept largely in terms of such an analysis. A conclusion is that in spite of more radical agricultural programmes and a larger volume of inputs in China, the agricultural production in India experienced a higher growth during the period 1952-73. The basic explanation of this paradox, the author argues, seems to lie partly in the relative levels of land productivity of the two countries during the early 1950s, and partly in the productive incentive of the peasantry. The latter element, a fundamentally stronger incentive for Indian peasants, is ascribed to the hasty industrialization policy in China, which left the initiative in the rural economy to the peasantry.

18. Pranab Bardhan: 'Agriculture in China and India. Output, Input and Prices'. (1969).

This study includes both an analysis of performance in agricultural production and a description of price policies in India and China during the period 1952-67. In addition the paper includes a discussion of the marketed surplus of food grain production in each country.

Performance in agricultural production is examined in terms of input-output analysis. As in the preceding study by Bandopadhyaya, the author arrives at the conclusion that India has experienced a higher growth rate than China, despite a lower level of inputs. The author argues that collectivisation and the limit for exploitation of traditional factors of production accounts for the comparative failure of Chinese growth.

In the discussion of the marketed surplus of food grain production, the author shows that the amount brought on the market has undergone a decline (in percentage) in both countries during the period described. It is suggested that this may be the result of less extractive policies towards the peasants in China, while, in India, price speculations on the part of the middle and large-scale farmers may have caused this decline.

19. Gilbert Etienne: 'Foodgrain Production and Population in Asia: China, India and Bangladesh'. (1977).

This study is devoted to the issue of population growth versus growth in food grain production in three heavily populated countries of Asia. Thus, it attempts to refute the widespread predictions of massive hunger in these countries by the end of the century. It deals initially with the expected growth of population in each nation, which is likely to decline as the family planning programme, organized on a considerable scale in both China and India, show their impact. Then, the prospects for food production are considered on the basis of a discussion of agrarian reforms and the trends in output of food grains experienced in the period 1952-75. An appendix lists the basic statistics for this discussion in eight tables.

The author arrives at the conclusion that the situation for these countries is by no means desperate, as the production of food grains as well as cash crops and sideline production is likely to increase beyond the growth of the population. However, a number of structural changes in the rural economy, while diminishing the effects of natural calamities, has also exposed agricultural development to human shortcomings and deficient planning. This is the case because agricultural modernization in India and China is critically dependent on modern inputs such as fertilizers and scientific research. Finally, future expansion in production will depend on the expansion and improvement of aspects of the rural infrastructure, for example, power supply and road networks.

20. Netra Pal Jain: Rural Reconstruction in India and China. A Comparative Study. (1970).

This volume describes the reorganizations and their effects in Indian and Chinese agriculture during the first two five-year plans. It is divided into two parts: part one describes organizational issues, focusing on community development in India and the rural People's Communes in China; part two surveys the performance of the agricultural economies and discusses inputs and the diversification of rural economy in each country.

In his conclusion Jain argues that through the land reforms both countries have been equipped with a base for an acceleration of agricultural production. In the period studied, India has achieved better results in overall growth in agricultural production, while China has been more successful in improving the mobilization of inputs for agricultural development and the diversification of the rural economy. Furthermore, the Chinese experience, although it has probably been based on a different social and political philosophy, may contain elements which can be transferred to India - or at least be regarded as a challenge to Indian development.

21. Jon Sigurdson: 'Development of Rural Areas in India and China'. (1976).

This paper contains a comparative discussion of approaches to rural development, treating a number of problems which are apparent in both India and China, notably, the questions of unemployment and the diversifications of the rural economy, and the marginalization issue.

The question of employment is pressing because of a predictable increase in the working-age population during the next decades in both countries. In addition, technological modernization of agriculture is likely to have an adverse effect on employment. The government programmes and social organizations established in India and China to deal with these problems are reviewed, and their capabilities and effectiveness evaluated.

Clustering of industrial activities around major cities is a predominant pattern in developing countries, a pattern that contributes to the marginalization of the rural economy. The author describes a number of policies adopted in China to counter this marginalization effect, and finds that similar efforts in India have had little success.

Finally the author considers the question of transfer of strategies from one country to the other. It is argued that it is important to distinguish policies and programmes which can be easily transferred to different socio-political contexts from those which are deeply embedded in their unique environments.

22. Jon Sigurdson: 'Water Policies in India and China'. (1977).

The water policies analysed in this paper include policies on irrigation and flood control, inland navigation, hydropower, and inland fishery. The first part describes the water resources in India and China and presents an outline of the Chinese and Indian strategies for developing these potentials. A basic difference is also found in the administrative structure of water control, for instance in relation to mobilization of manpower for construction projects.

In the second part of the paper, water policies and their implementation are analysed in detail. The author concludes that in many cases, the Chinese approaches - in particular the small-scale approach - have succeeded in tapping more extensively from the potential resources than the approaches adopted in India. This, the author argues, may be the result of a political philosophy which emphasizes decentralization, mass mobilization, and participation guided by national and community interests.

23. Jon Sigurdson: 'Rural Industrialization: A Comparison of Development Planning in China and India'. (1978).

This paper describes the considerable differences in rural industry promotion and development in India and China. The comparison is made on several points: (1) the objectives and organization of the development of rural industries, (2) the actual implementation of the development strategies, (3) the economic role of rural industrialization with respect to employment and contribution to the total industrial production value, and (4) the role of rural industrialization in the socialist transformation of the Chinese economy.

The comparison is based on a description of the policies adopted in each country, and a difference is noted in the objectives behind these policies: the Chinese planners stress skill formation and the development of regional industrial structures while employment *per se* is emphasized in India. It is arguec that the objective of employment generation has not been relevant in either the Indian or the Chinese reality. Consequently the questions of, for instance, skill formation and changes in urban-rural relationships are much more important in the developing economy. In addition it is argued that the organizational features of Chinese society may provide a better background for the rural industrialization programme, and that these features may be specific for China.

Finally the conception of rural industries as an instrument in the socialist transformation of Chinese society is discussed. The author finds that rural industrialization forms a part of the effort to increase the value of collective enterprise in relation to total production in rural areas. Of course this aspect of rural industrialization has no comparable equivalent in India, and this fact underscores the notion that development strategies may not be transferable in their total conceptions.

D. Education and Science

This section of the survey covers, unfortunately, only a few items. Research on education strategies and science policies on science, which should be of particular interest to the readers of this volume, has been in a deplorable state of affairs for a long time. Most of the items surveyed here are available in prepublication form only, and from a methodological point of view the descriptive approach adopted in the studies contributes little to general theoretical knowledge of the subject.

Some of the papers included in this volume, however, notably Aqueil Ahmad's paper on science policy in India and China, are intended to fill this gap. Moreover, it is the expressed ambition of the editors of this volume that it might contribute to an additional research effort in this area in the future.

24. Aqueil Ahmad: 'Flows of Science and Technology Information: The Cases of India and China'. (1978).

This paper is concerned with three issues: the basic structural features of the science and technology systems in India and China, the transfer of science and technology to society, and a discussion of the democratization of science.

In the first section it is argued that the flow of science and technology information in India is closely linked with the West, and that this constitutes simultaneously the strength and the weakness of the system. This contrasts with China, which has a self-reliant science and technology system despite substantial inflows from the West.

In the second section the science and technology systems are evaluated with reference to information flow from research institutes to all levels of society. The Chinese policies are seen as conducive to this flow, while a number of inhibiting factors are described in the case of India. Finally Chinese efforts in popularization of science and technology are compared with a number of programmes in India, notably the People's Science Movement in Kerala.

25. G.C. Beri: 'The Development of Education and Professional Manpower in China and India. A Comparative Study' I-II. (1969).

This study covers the period 1949-63 and is divided into two separate parts, dealing with the development of education and professional manpower respectively. Only part II was available for the present survey. It describes the magnitude and utilization of the following categories of manpower: engineering, medical and scientific personnel plus graduate teachers and agricultural graduates.

The description centres around a listing of figures for the quantitative growth of manpower in each nation, and contains no analysis of the factors which may have affected the divergent developments. The conclusion is reached that the growth of professional manpower - excepting scientific manpower - was greater in China, but doubts are expressed on the quality of the Chinese graduates.

26. Rushikesh Maru: 'Research and Development in India and China: A Comparative Analysis of Research Statistics and Research Effort. Preliminary Draft'. (1969).

This study is an attempt to compare the research and development efforts of India and China in quantitative terms for the period 1950-65. The comparison concentrates on two indicators of research and development in the two countries, the expenditures and manpower in research and development. Figures for each indicator are compared in four base years: 1950, 1955, 1960, and 1965.

The study of statistical figures for these two indicators in each country includes careful discussions of methodological problems, for instance the statistical system in China and the influence of divergent exchange rates. A summary of the conclusions, presented in Section V, indicates that in terms of the two indicators, India experienced a larger growth in the 1950s, but that China gained considerably in the period 1960-5.

27. Rushikesh Maru: 'Health Manpower Strategies for Rural Health Services. India and China: 1949-1975'. (1976).

This paper is a slightly revised version of a chapter in the author's dissertation on birth control in India and China (see item no. 32). The author focuses on the strategies pursued by the leadership of India and China with reference to health manpower, as well as the progress achieved in training health manpower in the two countries. The period analysed has been divided into two periods, before and after 1965 respectively. For each of these periods the questions of redistribution of health manpower from urban to rural areas and the utilization of indigenous medicine are discussed; in addition the formal training of health manpower has been described.

The author argues that in the period before 1965, both India and China pursued the 'professional' strategy, which emphasises advanced training in Western medicine and concentrates skilled medical manpower in urban areas. After 1965, however, China changed to a 'populist' strategy, which integrated short-term training of 'barefoot doctors' with advanced training and traditional medicine with Western medicine. India, on the other hand, appears to have continued a predominantly professional strategy.

28. Kristina Thuselius: 'Utbildningsplanering, teori och praktik. En jämförelse mellan Kina och Indien' (Educational planning, theory and practise. A comparison of India and China). (1975).

This short report presents an overview of the development of education in India and China. It also includes a discussion of the planning of education in each country, and the economic implications of education planning in developing countries.

The Chinese and Indian approaches are described in relation to a number of contradictions in educational planning, such as quality versus quantity and the importance of society versus the importance of the individual.

E. Technology, Health and Foreign Trade

Two studies deal with technology and technological choice in India and China; Aiyar compares Chinese and Indian industrial products, focusing primarily on the figures for size, capacity, etc. for the products. This method appears to put too much emphasis on the technical magnitude of the industrial products, and largely ignores the question of appropriateness. Bhalla, in contrast, compares the choice of technologies in construction in India and China, focusing on the Chinese experience of combining capital-intensive and labour-intensive technology according to circumstances.

The health services have been the subject of studies by Birke and Maru; Birke describes briefly the apparent effects of divergent health service strategies in India and China, while Maru examines political and economic factors in the development of population control programmes.

Finally, Ranjit Sau's paper on the foreign trade of India and China presents some interesting conclusions about the status of each country in regard to over- and underpricing of commodities.

From the methodological point of view, only Bhalla and Maru seem to have developed analyses beyond the descriptive stage. Both have tried to evaluate the effects of political, economic and social factors in the development of strategies and the implementation of policies, thus attempting to isolate the elements in development which are subject to political manipulation.

29. Swaminathan S. Aiyar: 'India Ahead of China in Some Fields: Danger of Falling Behind'. (1978).

This short paper summarizes a report made by the Association of Indian Engineering Industry which compares Chinese products in selected areas with similar Indian products.

The report concludes that Indian products are superior to Chinese products in some fields, notably electric motors and cement plants, while China has a lead over India in other fields, for instance railway equipment. The criteria for determining the lead appear to be figures for the size, capacity, etc. of the products. It remains doubtful whether this method of evaluating technological progress, which completely ignores the question of how appropriate the products are in their utilization contexts, is sufficient for an assessment of industrial development.

30. A.S. Bhalla: 'Technological Choice in Construction in Two Asian Countries: China and India'. (1974).

In this paper, Bhalla studies the influence of factor-price considerations in technological choice in capital construction projects. The author studies the Chinese experience in implementing the 'two-leg' strategy in three phases: the Soviet model phase (1949-57), the Great Leap Forward phase (1958-60), and the sixties (1961-70). In these periods the Chinese have experimented with capital- and labour-intensive technologies, as well as with large-scale labour-mobilization programmes.

In a comparison with the Indian experience, the author argues that planning and organizational factors largely determined technical choices in construction, and that reasons 'beyond factor prices' were important even in a market-oriented economy. Furthermore labour mobilization was more successfully undertaken in China, partly because of better organization and decentralized planning at the local level.

31. Lynda Birke: 'India and China'. (1978).
A short article devoted to a cursory, generalized comparison of health care in India and China. The first section relates some personal experiences from visits to India and China. One is 'a picture of hopelessness' for the majority of the people (India), while the other is 'full of hope' (China).

The next sections compare Indian and Chinese efforts to integrate traditional medicine into medical care, to provide widespread medical education and extend health services to the rural areas, and finally, to curb the influence of multi-national corporations in the pharmaceutical services. The Chinese achievements are regarded as superior, and 'the consequences are, quite simply, better health for the Chinese people, and a greater life expectancy'.

32. Rushikesh Maru: Birth Control in India and in the People's Republic of China. (1976).
This doctoral dissertation is an examination of birth control programmes and their implementation in India and China. An empirical analysis includes descriptions of policies on birth control in each country, the contraceptive technologies applied, and the organization of family planning and health services in India and China. In addition, Maru describes the actual implementation of family planning policies in both countries.

The author concludes that despite political differences, both countries responded to the population problem by instituting national birth-control programmes. Furthermore, economic or other factors were more important than political factors influencing programme priorities, choice of technology and production of contraceptives. In contrast, organization and implementation of policies on family planning depended to a large degree on such factors as organizational capabilities and the permeability of the system to foreign influence.

33. Ranjit Sau: 'Some Aspects of Foreign Trade of India and China'. (1976).

This paper examines the price differentials for import and export commodities charged by various countries engaged in trade with India and China. In so doing the author attempts to elucidate trends in overpricing and underpricing in trade between Third World countries and industrialized countries in particular, but also among Third World countries themselves.

By means of a rather complicated computational method, the author arrives at figures for the relative price index for exports and imports to and from some 40 countries trading with India and China. This index can roughly be said to show the ratio between the price charged by a particular country for a certain commodity and the 'average world price' for the same commodity. Although the author abstains from presenting conclusive statements about over- and underpricing, a glance at the tables show that great divergences exist between imports and exports for each country, and among the countries included in the analysis. The indexes are computed for two periods in the case of India (1968-9 and 1970-1) and four base years in the case of China (1964, 1966, 1968, 1970). This gives the possibility of comparing trends over a limited period of time for each country.

The paper, however, contains very few comparative statements as such, and the author explicitly warns against sweeping conclusions on the basis of these indexes.

THE METHODOLOGY OF COMPARATIVE RESEARCH: A SELECTED BIBLIOGRAPHY

Erik Baark

The methodology of comparative research has been subject to a number of theoretical discussions during the last three decades. Therefore, a substantial body of literature exists which deals with the problems encountered in comparative research. In this selected bibliography, a few of these works are reviewed; they represent some of the theoretical literature consulted in the process of writing chapter I, and the annotations are thus designed to indicate which papers were found relevant to India-China comparative research.

One point, however, is implicit in the majority of the theoretical discussions presented in these works: more than two nations are compared. Often a comparison of virtually all the countries in the world is implied. This assumption has the effect of focusing the discussions on problems of worldwide comparisons, problems which may be absent or of minor importance in comparisons dealing with two countries. For instance, questions concerning the collection of statistics from a large number of nations, establishment of data-banks for cross-cultural indicators and the establishment of typologies tend to focus the discussions on quantitative measures and indicators. With the present status of aggregate statistics in China, comparative research involving India and China has, on the contrary, tended to employ qualitative data. There is thus a discrepancy between the problems encountered in comparative research generally and the problems pertinent to India-China comparative research in particular.

Another important difference is that the theoretical discussions on methodology presented in this bibliography focus on political science (i.e. comparative politics), sociology and economics. The present volume is concerned with development studies, or rather, on the role of technology and science in development. Development research has flourished as a separate research field only in very recent times, and research into development issues has not - to my knowledge - developed a methodology of its own. Thus development research has largely borrowed its methodological approaches primarily from the disciplines mentioned above, political science, sociology and economics. This pattern will probably continue until a manifest need for a specialized methodology for development research is clearly recognized within academic circles.

1. Douglas E. Ashford, (ed.), *Comparing Public Policies: New Concepts and Methods,* London, 1978.

This volume is, as the title suggests, concerned with the comparison of public policies, mainly in industrialized countries. Two papers, however, deal with general theoretical approaches to comparative research. These are 'The Logic of Public Problems: Evaluation in Comparative Policy Research', by C.W. Anderson and 'A Logic of Comparative Policy Analysis', by Henry Teune. Anderson explores the problems experienced by researchers in defining the scope and focus of research in comparative politics. He suggests that at least some personal judgement is necessary in the process and because political evaluation is often the explicit goal of political research, a normative analysis may contribute more to a knowledge of the rationale behind statecraft than does 'pure, deductive conceptual analysis'. Henry Teune examines the problems of improving the predictions of the impact of particular policies. He argues that policy predictions can be improved by: first, the direction of theoretical concerns to macropolicy analysis and point predictions, and second, the use of a 'most similar system' design in empirical research. In addition, he describes the logic of a 'quasi-experimental design' in empirical research.

2. Harry Eckstein and David E. Apter, (ed.), *Comparative Politics: A Reader,* London, 1963.

This book reflects the state of the art in comparative politics at the beginning of the 1960s. It is intended as a collection of reference material for students of comparative government, and contains excerpts from various books and articles on that subject. A paper by Gunnar Hecksher, 'General Methodological Problems', is interesting, as it discusses many of the problems and methodological questions involved in both comparative politics and in humanistic studies in general. Among other instructive comments, it contains the following definition of methodology:

> What 'methodology' is can thus be stated rather simply: to take stock of methods used so far, criticize them on the basis of the results attained in employing them and trying to perfect them further with a view to the future. This is something which we are all doing, consciously or un-

consciously. But it is particularly for countries where the study of political science is just beginning that we need to present explicitly our experiences.

3. Robert T. Holt and John E. Turner, (ed.), *The Methodology of Comparative Research,* New York, 1970.

This volume is addressed to the theoretical questions of methodology in comparative research in the social and political sciences. It is specifically concerned with paradigms in comparative politics, but papers are also included which deal with practical procedures in comparative research, such as the cross-cultural use of survey-research techniques and the use of historical materials in cross-cultural studies. An introductory chapter by the editors is of particular interest. It deals with comparative methodology in general terms, and describes the application of the scientific method (which is called the analytic method in chapter I). A number of issues relevant to this method are discussed, for instance the control of background factors by randomization and specification, the operationalization of concepts and problems of comparability of sampling.

4. Lawrence C. Mayer, *Comparative Political Inquiry. A Methodological Survey,* Homewood, Illinois, 1972.

In Mayer's volume on comparative methodology, the concept of the scientific method is described in great detail. The influence of modern writings on the philosophy of science, notably Thomas Kuhn's concept of a scientific paradigm, is explicitly acknowledged, and the ideal scientific method defined in the first part of the book is used to evaluate the methodologies of different empirical and theoretical studies in comparative politics. Part I, dealing with epistemology, is thus the most valuable section of the book from the viewpoint of methodological theory. There is also a glossary of terms employed in the field of comparative politics and methodology, plus bibliographical notes listing and commenting briefly on literature on the subject.

5. Stein Rokkan, (ed.), *Comparative Research across Cultures and Nations,* Paris, 1968.

This book presents papers from a conference on comparative research held in Paris 1965. In addition to an introduction which describes the efforts of the International Social Science Council in the field of comparative research, the papers are presented under three headings: 'Cross-cultural comparisons', 'Comparative histories of processes of development', and 'Quantitative approaches to cross-national comparisons'. The papers collected in part 2 are particularly interesting to the readers of the present volume. A paper on the comparative analysis of processes of modernization deals with the political and economic background of modernization in general terms; another paper on the empirical and statistical basis for comparative analyses of historical change discusses the accumulation, standardization, and evaluation of information for comparative research. The International Social Science Council appear to be in the process of developing a cross-national databank of information (mostly quantitative) for comparative research.

6. Howard A. Scarrow, *Comparative Political Analysis: An Introduction,* New York, 1969.

This book resembles Mayer's book on comparative political inquiry (item 4), as it is also a presentation of methodological approaches and an evaluation of the literature published in the field of comparative politics. It differs by recognizing several 'logics' in methodology, such as descriptive and explanatory modes, functional and casual explanation, etc., while Mayer defines only one method, the 'scientific method'. Therefore Scarrow's discussion of the difference and relation between description and explanation has been particularly valuable for the establishment of the concepts of descriptive research methods in chapter I of this volume.

7. Donald P. Warwick and Samuel Osherson, (ed.), *Comparative Research Methods,* Englewood Cliffs, N.J., 1973.

This volume contains a selection of articles which were used at a graduate seminar on comparative international studies. The selection of articles and the focus of the discussions thus show a definite bias towards sociological research. In addition to a paper by Niel J. Smelser on the methodology of comparative analysis, which deals with scientific inquiry and levels of comparative analysis, the volume contains a number of interesting articles on sampling techniques in cross-cultural research.

BIBLIOGRAPHY OF INDIA-CHINA COMPARISONS

1. AHMAD, Aqueil
'Flows of Science and Technology Information: The Cases of India and China' (Paper prepared for the Conference on Communication Support for Rural Development, Administrative Staff College of India, Hyderabad, 27-30 December 1978), Hyderabad 1978.
2. AHMAD, Aqueil
'Science and Technology in Development. Policy Options for India and China', *Economic and Political Weekly* Vol. XIII, Nos. 51 and 52, December 1978, pp. 2079-2090.
3. AIYAR, Swaminathan S.
'India Ahead of China in Some Fields. Danger of Falling Behind', *Times of India,* 22 September 1978.
4. ANDERSON, Mary Boughman
'Self-reliant Development: A Comparison of the Economic Development Strategies of Mohandas Gandhi, Mao Tse Tung and Julius Nyerere' (PhD Thesis, University of Colorado: Dept. of Economics), 1978.
5. BANDYOPADHYAYA, J.
Mao Tse-tung and Gandhi, New Delhi, 1973.
6. BANDOPADHYAYA, Kalyani
Agricultural Development in China and India: A Comparative Study, New Delhi, 1976.
7. BARDHAN, Pranab
'Agriculture in China and India. Output, Input and Prices', *Economic and Political Weekly* Vol. IV, Nos. 1 and 2, January 1969, pp. 53-65.
8. BERGMANN, Theodor
The Development Models of India, the Soviet Union and China: A Comparative Analysis, Amsterdam 1977.
9. BERI, G.C.
'The Developing of Educational and Professional Manpower in China and India: A Comparative Study - I' *China Report V* No. 4, July-August 1969, pp. 1-8.
10. BERI, G.C.
'The Developing of Educational and Professional Manpower in China and India: A Comparative Study - II' *China Report V* No. 5, September-October 1969, pp. 9-13.
11. BHALLA, A.S.
'Technological Choice in Construction in Two Asian Countries: China and India', *World Development* Vol. 2, No. 3, March 1974, pp. 65-73.

12. BHATTACHARYA, Debesh
'India and China - Contrast and Comparison 1950-72' *Journal of Contemporary Asia* Vol. 4, No. 4, 1974 pp. 439-59.
13. BIRKE, Lynda
'India and China' *China Now* No. 81, November/ December 1978, pp. 5-7.
14. BYRES, T.J. and NOLAN, Peter
Inequality: India and China Compared 1950-70, Milton Keynes, 1976.
15. CHEN, Kuan-I and UPPAL, J.S.
Comparative Development of India and China, New York, 1971.
16. ETIENNE, Gilbert
'Foodgrain Production and Population in Asia: China, India and Bangladesh' *World Development* Vol. 5, Nos. 5-7, 1977.
17. FUKUTAKE, Tadashi
Asian Rural Society: China, India, Japan, Tokyo, 1967.
18. HARRIS, Nigel
India-China. Underdevelopment and Revolution, New Delhi, 1974.
19. HETTNE, Björn
'Energipolitiken i Indien och Kina. Beroende Kontra Autonomi' Vetenskapsakademiens energikonferens i Lerum, 27-31 October 1975.
20. HETTNE, Björn
Utviklingsstrategier i Kina och Indien, Lund, 1979.
21. JAIN, Netra Pal
Rural Reconstruction in India and China, New Delhi, 1970.
22. LATEEF, Sarwar
Economic Growth in China and India, 1950-1980, Vols. 1 and 2, London, 1976.
23. LIU, Ta-Chung and YEH, Kung-chia
'Chinese and Other Asian Economies: A Quantitative Evaluation' *American Economic Review,* May 1973.
24. MACRAE, F.T.
'India and China - A Comparative View for the 1970s' *Coexistence* No. 8, 1971, pp. 41-48.
25. MAKHIJANI, Arjun and POOLE, Alan
Energy and Agriculture in the Third World, Cambridge, Mass., 1975.
26. MALENBAUM, Wilfred
'India and China: Development Contrasts' *The Journal of Political Economy,* Feb. 1956.

27. MALENBAUM, Wilfred
'India and China: Contrasts in Development Performance' *American Economic Review,* June 1959, pp. 284-309.
28. MALENBAUM, Wilfred
'Growth Models of Asia: China or India?' *International Journal,* Spring 1960, pp. 138-143.
29. MARU, Rushikesh
'Research and Development in India and China: A Comparative Analysis of Research Statistics and Research Effort. Preliminary Draft' Unpublished paper, Research Policy Program, Lund, 1969.
30. MARU, Rushikesh
'India and China: Towards Self-reliant Economies' *China Report,* Dec. 1965-Jan. 1966, pp. 11-15.
31. MARU, Rushikesh
'Health Manpower Strategies for Rural Health Services - India and China, 1949-1975' *Economic and Political Weekly* Vol. XI, Nos. 31-33, August 1976, pp. 1253-68.
32. MARU, Rushikesh
Birth Control in India and in the People's Republic of China, Ahmedabad: Population Project Unit, Indian Institute of Management, November 1976.
33. MURPHY, Rhoads
The Outsiders: The Western Experience in India and China, Ann Arbor, 1977.
34. PATRICK, Hugh and Peter SCHRAN
'Economic Contrasts: China, India and Japan' *Journal of International Studies* Vol. XVII, No. 2, 1963, pp. 168-184.
35. RAJ, K.N.
India, Pakistan and China: Economic Growth and Outlook, Bombay, 1967.
36. RAO, K.R.M.
'India Measured Against China' *Democratic World* No. 7, Nov. 1976, pp. 3-4.
37. RICHMAN, Barry
'Economic Development in China and India: Some Conditioning Factors' *Pacific Affairs* Vol. 45, No. 1, Spring 1972, pp. 75-91.
38. RICHMAN, Barry
'Chinese and Indian Development: An Inter-disciplinary Experimental Analysis' *American Economic Review* Vol. 65, No. 2, May 1975, pp. 345-355.

39. SAU, Ranjit
'Some Aspects of Foreign Trade of India and China' *Economic and Political Weekly* Vol. XI, Nos. 31, 32 and 33, August 1976, pp. 1315-20.
40. SETHI, J.D.
'Mao and Gandhi: Convergence and Divergence - A Preliminary Note' *China Report* Vol. XIII, No. 1, January-February 1977, pp. 16-33.
41. SIGURDSON, Jon
'Development of Rural Areas in India and China' *AMBIO* (Royal Swedish Academy of Science Journal), Vol. 5, No. 3, 1976, pp. 89-107.
42. SIGURDSON, Jon
'Water Policies in India and China' *AMBIO* Vol. 6, No. 1, 1977, pp. 70-76.
43. SIGURDSON, Jon
'Rural Industrialization: A Comparison of Development Planning in China and India' *World Development* Vol. 6, May 1978, pp. 667-680.
44. SINHA, R.P.
'Competing ideology and agricultural strategy: Current agricultural development in India and China compared with Meiji strategy' (Paper delivered at the Third European Conference on Modern South Asian Studies, Heidelberg), 1972.
45. SWAMY, Subramaniam
'Economic Growth in China and India, 1952-1970: A Comparative Appraisal' *Economic Development and Cultural Change* Vol. 21, No. 4, Part II, July 1973, pp. 1-80.
46. SWAMY, Subramaniam
'The Economic Distance between China and India' *China Quarterly* No. 70, June 1977, pp. 371-82.
47. THUSELIUS, Kristina
'Utbildningsplanering, teori och praktik. En jämförelse mellan Kina och Indien' Report, Research Policy Program, Lund 1975.
48. VAUGHAN, Patricia A.
'Science and Science Policy in India and China: A Comparative Approach' (Thesis submitted to the Department of Political Science, University of Melbourne), 1976.
49. VAITHESWARAN, R.
'Gandhi and Mao: A Comparison in Terms of Relevance for the Politics of National Liberation and Reconstruction' *China Report* Vol. XII, Nos. 5 and 6, September-December 1976, pp. 31-43.

50. WEISSKOPF, Thomas E.
'China and India. A Comparative Survey of Performance in Economic Development' *Economic and Political Weekly* Vol. X, Nos. 5, 6, and 7, February 1975, pp. 175-194.
51. WEISSKOPF, Thomas E.
'China and India: A Comparative Survey of Economic Development Performance' *American Economic Review* Vol. 65, May 1975, pp. 345-71.

APPENDIX I
LIST OF PARTICIPANTS AT THE SYMPOSIUM 'TECHNOLOGY IN DEVELOPMENT: THE CASES OF INDIA AND CHINA'

Aqueil AHMAD, Research Policy Institute (RPI), Lund, and Administrative Staff College of India, Hyderabad, India.

G.S. AURORA, North Eastern Hill University, Shillong, India.

Erik BAARK, RPI, Lund.

Boel BILLGREN, RPI, Lund.

Aadel BRUN TSCHUDI, Department of Geography, University of Oslo, Oslo.

Richard CONROY, Science Policy Research Unit, University of Sussex, Brighton, England.

Stefan DEDIJER, RPI, Lund.

Charles EDQUIST, RPI, Lund.

Olle EDQUIST, Swedish Agency for Research Cooperation with Developing Countries (SAREC), Stockholm.

Aant ELZINGA, Theory of Science, University of Gothenburg, Gothenburg

Jeremy GOULD, Institute of Development Studies, University of Helsinki, Helsinki.

Amalendu GUHA, Oslo.

Kjell HAVNEVIK, Chr. Michelsens Institute, Bergen.

Karl Reinhold HAELLQUIST, Scandinavian Institute of Asian Studies, Copenhagen.

Michael HAMMARSKJÖLD, Department of Human Geography, University of Lund, Lund.

Jan HESSELBERG, Department of Geography, University of Oslo, Oslo.

Björn HETTNE, Department of Peace and Conflict Research, University of Gothenburg, Gothenburg.

Robert C. HSU, RPI, Lund, and Clark University, Worcester, USA.

Gerhard HULTCRANTZ, Stockholm School of Economics, Stockholm.

Göran HÖGBLOM, Swedish University of Agricultural Sciences, Uppsala.

Joachim ISRAEL, Department of Sociology, University of Lund, Lund.

Andrew JAMISON, RPI, Lund.

Anders KARLQVIST, Delegation for Future Oriented Studies, Stockholm.

Kjeld Allan LARSEN, Copenhagen.

Theo LEHNER, Swedish Metallurgical Research Institute, (MEFOS), Lulea.

APPENDIX I

Hakan LINDHOFF, Stockholm School of Economics, Stockholm.
Asbjørn LØVBRAEK, Peace Research Institute of Oslo, Oslo.
Ward MOREHOUSE, RPI, Lund, and Council of International and Public Affairs, New York, USA.
Poul MOHR, Scandinavian Institute of Asian Studies, Copenhagen.
Peter NEERSØ, Roskilde University Centre, Roskilde.
Kyösti PULLIAINEN, University of Joensuu, Joensuu.
Jairam RAMESH, Office of Science and Technology, The World Bank, Washington, D.C., USA.
Torbjörn SANDSTRÖM, Chalmers' Institute of Technology, Gothenburg.
Jon SIGURDSON, Scandinavian Institute of Asian Studies, Copenhagen, and RPI, Lund.
Uno SVEDIN, Secretariat for Future Studies, Stockholm.
Mats THORÉN, Academy of Engineering Sciences, Stockholm.
Marina THORBORG, University of Uppsala, Uppsala.
Donald WAGNER, Scandinavian Institute of Asian Studies, Copenhagen.
Jan VALDELIN, Stockholm School of Economics, Stockholm.
Gustaf ÖSTBERG, Lund Institute of Technology, Lund.

APPENDIX II
LIST OF PAPERS PRESENTED AT THE SYMPOSIUM 'TECHNOLOGY IN DEVELOPMENT: THE CASES OF INDIA AND CHINA'

Aqueil Ahmad: 'Science and Technology Policy Options: India and China'.

Richard Conroy: 'The Growth and Development of the Instruments Sector in China and India - A Comparison'.

Aant Elzinga: 'The Chinese Science Policy Debate 1972-1976'.

Amalendu Guha: 'Two Development Models and Strategies Compared: China and India'.

Björn Hettne: 'Self-reliance Versus Modernization - The Dialectics of Development Strategies in India and China'.

Robert Hsu: 'Agricultural Mechanization in China'.

Göran Högblom: 'Some Views of the Organisation of Agricultural Research and Extension in the PRC'.

Ward Morehouse: 'Local Technology Systems in India and China: Prolegomonon to Comparative Analysis of Technological Change in the Alleviation of Poverty'.

Peter Neersø: 'Some Aspects of India's Policies on the Import of Technology'.

Marina Thorborg: 'Technological Change and Women Participation in Production in the PRC'.

For Product Safety Concerns and Information please contact our EU representative GPSR@taylorandfrancis.com
Taylor & Francis Verlag GmbH, Kaufingerstraße 24, 80331 München, Germany

www.ingramcontent.com/pod-product-compliance
Lightning Source LLC
LaVergne TN
LVHW020638300926
849161LV00004B/740